D1153150

Beagle

The European Space Agency Mars Express and Beagle 2 mission
Artist's impression of the approach to Mars
Courtesy ESA

Beagle

from sailing ship to Mars spacecraft

Colin Pillinger

faber and faber

First published in 2003
by Faber and Faber Limited
3 Queen Square, London WC1N 3AU

Printed in Malta by Gutenberg Press

A CIP record for this book is available from the British Library

ISBN 0-571-22323-0

10 9 8 7 6 5 4 3 2 1

Contents

To a finish

Preface

The name Beagle 2 was chosen for the spacecraft that carries Britain's hopes to discover life on Mars by my wife Judith during a journey back from Paris in April 1997. I had just attended the European Space Agency's meeting to discuss the possibility of a mission to the red planet called Mars Express and had suggested the project should include a lander. I needed a landing spacecraft to see whether work carried out on meteorites from Mars was correct in predicting that life on a second planet in the solar system was possible; the meteorite samples might even have already revealed its presence but for such a momentous claim we had to be sure. The experiments had to be repeated on Mars.

Only hours earlier I had been asked who would pay for a martian lander and who would develop it? It was pretty clear that ESA did not have the resources and they thought no one else did either. I had said it was too good a chance to miss, someone will pay. If I was going to raise the funds from sources other than the usual ones we realised we would need a name that was instantly recognisable. HMS *Beagle* was the ship that had taken Charles Darwin around the world and led to his writing *On the Origin of Species*; it is one of the most well-known ship names. Her voyage led to the discovery of the secret of life on Earth, could we do the same for Mars? We decided then and there to honour the ship, the vehicle which made Darwin's contributions possible. So Beagle it would be for the spacecraft.

It was not until later in September of that year, whilst preparing for a presentation to ESA, outlining the evolving Beagle 2 lander project, that I first pointed out the similarities between our situation and the opposition that Darwin initially faced from his father. I also informed the audience that financing the voyage of HMS *Beagle* had not been all plain sailing either as Captain FitzRoy knew to his cost.

The following March at the Geological Society I expanded on the Beagle 2 and HMS *Beagle* similarities. I quoted from the texts preached at the sermons at Darwin's baptism and at his funeral and

"where there is no vision, the people perish"
(Proverbs Chapter 29 v18)

had become the dedication in a proposal submitted to ESA. When the proposal was reviewed, and the project had to halve the size of the lander, there was the precedent of Darwin removing the drawer of his bureau to accommodate his feet whilst sleeping on the *Beagle* that I could refer to.

As the Beagle 2 project progressed more and more analogies with HMS *Beagle* were found until the point where it was recognised that there was a great opportunity for an exhibition at the National Maritime Museum in Greenwich, comparing and contrasting the pair. As preparations went ahead it became obvious that far more material was available than there was room for.

The exhibition evolved into a book and here is the product - the story of how it has come to pass that Beagle 2 is going to Mars juxtaposed against a nautical history of the name. It is quite remarkable how little has changed in terms of exploring a planet, whether it be Earth or Mars, in 170 odd years.

There are a few idiosyncrasies concerning this book which are entirely mine. I like to use the adjective martian with a lower case m to distinguish mentions of the planet from mythical inhabitants who I call Martians with a capital M. I'm told that spacecraft names, like those of ships, should be italicised but I have chosen this form for the ship only, using normal font for the spacecraft so that it is more easy to distinguish in the text.

Beagle 2 grew from an idea on a French motorway to a multinational enterprise. I have counted thirteen countries contributing but everyone will recognise it is a quintessentially British project. If I tried to acknowledge everyone involved in Beagle 2, I would miss someone out and that would be a tragedy. There are however a few who must be named: Sir John Daniel, Vice-Chancellor of The Open University at the start of the Beagle 2 venture who after listening to me for thirty minutes said "I think we might be able to pull this off", he delegated the pulling to his long-suffering Finance Director Miles Hedges; Paul Murdin of the British National Space Centre who said "I can't tell you to go ahead but then I can't tell you not to"; Mike Rickett of Matra

Marconi Space (now Astrium) who recognised the opportunities available for the UK space industry and appointed Jim Clemmet to engineer them; Lord Sainsbury who reiterated the party line "there is no money for this" at his first press conference as Science Minister, he then found rather a lot down the back of someone's sofa. And finally Mark Sims of the University of Leicester who had already sent a letter pleading to be allowed to join when I phoned to ask him if he would. He must have had second thoughts later because in his response to the circular announcing the first meeting and asking for expressions of interest he only ticked the 'maybe' box. He did come and has probably regretted it ever since.

To everyone else out there, 'the boys in the band', actually and metaphorically, no matter how big or small your contribution - thanks. We are on our way to Mars and no one can take that away from us.

Neither the book nor my involvement in the project would exist without the persistence of my wife. To her should go the credit, the mistakes are mine.

Colin Pillinger
September 2003

Acknowledgements

It is not possible to produce a book like this without the help and assistance of lots of people. Whilst I cannot namecheck the absolute multitude of contributors to the project itself for fear of missing someone, I am adopting the same approach to all those who helped and provided pictures, new material, offered and contributed information and directed us to sources, all of whom are gratefully acknowledged.

The magic of Beagle 2 was that everyone wanted to help so to all those of you who helped with the book, many thanks.

The source of images is acknowledged in picture captions. Whilst every effort has been made to contact and obtain permission from holders of copyright, if any involuntary infringement has occurred, sincere apologies are offered.

On the origin of Beagle

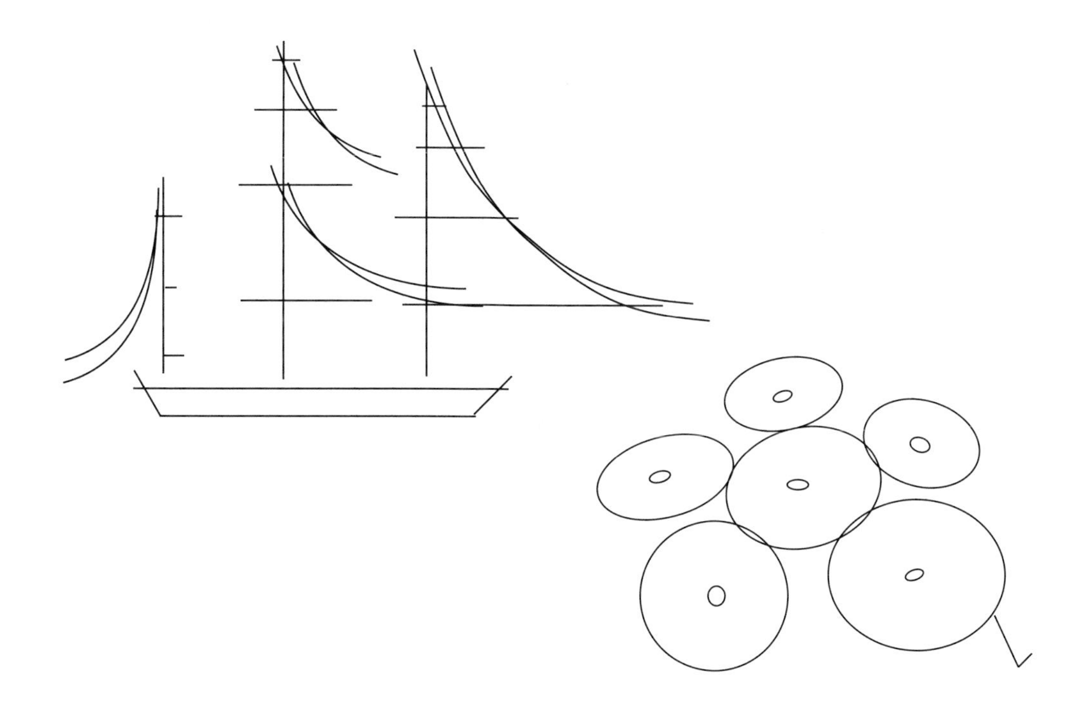

The spacecraft Beagle 2 is named in honour
of the ship captained by Robert FitzRoy which took
Charles Darwin around the world in the 1830s
and led to the writing of *On the Origin of Species*.

What's in a name

"Up to now the most important mission ever mounted in respect of the origin of life was Charles Darwin's voyage in HMS *Beagle*. We have therefore christened the concept Beagle 2."
Colin Pillinger
at the first meeting of what would become the British Mars lander team at The Royal Society, May 20th 1997.

Beagle 2 is the tiny British spacecraft which will land on Mars at the end of 2003. A part of the European Space Agency's Mars Express mission, its role is to search out evidence for life, past or present, on the red planet.

The project is named in honour of the ship captained by Robert FitzRoy which took Charles Darwin around the world in the 1830s and led to the writing of *On the Origin of Species*. So when it came to testing whether the theory of evolution extended to a second planet in the solar system, it seemed obvious that the spacecraft should be Beagle too. But since the Royal Navy already had an HMS *Beagle*, the space project became Beagle 2.

It turned out, however, that Beagle 2 was not just inheriting a name; the ship of the 1830s and today's lander are not so very different. Both represent state-of-the-art in exploration for their respective generations, leading the way in science and technology, to say nothing of adventure.

But the HMS *Beagle* of FitzRoy and Darwin is not the only naval ship proudly to bear the name. In all the British Navy have had nine HMS *Beagles*, many of which have had their moments of fame when they made news.

The name first came into use at the end of the eighteenth century, coincidentally it was just at that time when the media first recognised the public's fascination with the possibility of life in space. Beagle 2 owes its very existence to the lure of extraterrestrial life.

HMS Beagle *in Sydney Harbour by Owen Stanley FRS, c. 1841. Courtesy National Maritime Museum, London.*

Artist's impression of the Beagle 2 lander on the surface of Mars.

The first part of the story of *Beagle, from sailing ship to Mars spacecraft*, chronologically charts the unfolding of our understanding – and otherwise – of life elsewhere in the solar system, and in particular on Mars, following progress alongside a history of the various naval vessels.

The second half of the book continues with a detailed examination of the many parallels between Beagle 2 and its most illustrious predecessor, the third HMS *Beagle*, comparing one great age of discovery with what could be the next.

Radio Interviewer (during live broadcast)
"Why have you called the spacecraft Beagle 2?"
Author "In honour of HMS *Beagle* which was the ship which took Charles Darwin around the world and led to the understanding of evolution"
Interviewer "Were you part of that mission too?"
Author "I'm old but not that old!"

Journey of a lifetime

Mars Express and the Beagle 2 probe were launched from the Baikonur Cosmodrome in Kazakhstan at 11.45 pm local time on June 2nd 2003, by a Soyuz rocket with a Fregat upper stage. After a journey of nearly seven months, and a few days before reaching Mars, the probe, with lander inside, will be detached from the orbiter using a spin up and eject mechanism which will send it unpowered, but spinning for stability, towards the martian atmosphere. A real autonomous spacecraft for the first time, the final coast to Mars will begin; the final descent through the atmosphere on Christmas day will take a matter of minutes.

Slowed down by a heatshield, pilot and main parachutes, the lander, cocooned in gas-filled bags, will bounce to rest on the

Beagle 2 searching for signs of life on Mars

The mole collects samples from beneath the surface or under boulders and the corer grinder obtains material from within rocks. All specimens are analysed on the planet by the Beagle 2 on-board laboratory.

surface of Mars, at its destination on Isidis Planitia. Alone at last, Beagle 2 will open its lid, the solar arrays will fold out and a precious cargo of instruments and sample handling tools will begin the search for life. The package has the ability to conduct geochemical, geophysical, mineralogical, petrological, atmospheric, climatic, and environmental investigations, even astronomical studies, to understand the landing site in support of its main aim. The project is scheduled to last for 180 Earth days which is 175 and a half martian days.

Beagle 2 – first off the ground

Whilst Beagle 2 will be the first mission in space to carry the name, two other missions used the name Beagle, one was fictional and the other came to nought. Science fiction writer A.E. van Vogt called his 1950 Star Trek-style book *The Voyage of the Space Beagle*. Travelling from galaxy to galaxy its crew visited a number of planets and encountered the remains of past civilisations. They found that life forms existed in space itself.

Back in 1963, NASA initiated a study by the General Electric Company of a mission to land on Mars to search for life; it was intended to be a continuation of the Voyager programme using Saturn V rockets. To avoid confusion with projects already under investigation the code name Beagle was suggested. The study report says it was derived from the ship which took Charles Darwin around the world.

All resemblance to Beagle 2 ends there. NASA's project Beagle was to have two landers each 9,500 kilograms with a payload of 2,250 kilograms. Entry was to be behind an aeroshell of six metres diameter and when deployed the petals would measure ten metres across. And the cost, over one billion dollars at 1964 prices; a massive enterprise on any scale. For comparison the Beagle 2 probe is sixty-eight kilograms, with the lander weighing just thirty-three, its diameter is the size of a bicycle wheel and its cost at the outset was estimated as £25 million.

Not surprisingly the NASA Beagle never got off the ground.

One ship did however sail to Mars, in the children's fantasy written and illustrated by William Timlin in 1923.

The story opens with "Although it was difficult to believe, the Old Man had not always been old, and in his dim, forgotten youth, he had said 'I will go to Mars' But those around him, Scientists and Astronomers, some cried out in scorn."

Beagle 2 – the lander is the diameter of a bicycle wheel.

Naming ships

Sailors have always thought it necessary to name their ships. For example William the Conqueror's flagship from 1066 was *Moira*; Columbus journeyed to the new world with the *Nina*, *Pinto* and *Santa Maria* whilst John Cabot and his son Sebastian first landed on mainland America from *The Matthew*, whose crew numbered just seventeen souls.

The practice of handing down names from generation to generation began in the Tudor era. Sometimes a name which was first used in the 1500s has now been used twenty or more times.

A naval vessel that loses its commission is stripped of its name as well. The final approval for naming any ship of the line rests with the Sovereign; Charles II was responsible for introducing *Mars* to the list after the Restoration in the 1660s –the name of the God of War has been most frequently associated with battleships.

A replica of the name plate of the ninth HMS Beagle *presented by her Captain to the Beagle 2 team.*

Nothing is known about how *Beagle* became a ship's name. Beagle 2 though received royal recognition when the Queen, on a Jubilee sortie in 2002, offered support to the team saying "I am pleased that preparations for the Beagle 2 mission to Mars are well underway"

The Ship that Sailed to Mars, by William Timlin.
"Therefore he had taken his leave of men, and men's ways, and he had spent his long lifetime in a sleepy office in a dull, dark street; passing his waking hours in strange dreams, or poring over weird and ancient books, and always and ever planning a ship to sail to Mars.
By reason of his faith, the Fairies came to him, and he chose those with cunning skill as craftsmen to help him and started a shipyard".

Naming spacecraft

Who could ever forget the name of the World's first spacecraft? Sputnik, translated into English it means *travelling companion.*

For their manned programme the Russians followed up with Vostok (East) and, although they led the space race until the Moon landings, as a name it will never be remembered as well as the American Apollo programme. Non-manned missions heading out into the solar system did have quite romantic names such as Pioneer, Mariner and Voyager. All convey the sense of exploration but none of these quite reflects the spirit of adventure and achievement from times past.

In contrast to the old Soviets and NASA, the European Space Agency (ESA) has tended to honour famous people by naming space missions after them. For the mission to Halley's comet, ESA chose Giotto, the early renaissance artist who painted an apparition of the comet as part of a fresco on the wall of the Scrovegni Chapel at Padua. Darwin has been selected as the name for a multiple mirror telescope designed to be able to detect other planetary systems, analysing their atmospheres remotely for the purpose of discovering far away life, but this mission will not happen for many years yet.

Although Mars has long fascinated humankind, it seems not enough to inspire space mission engineers to come up with imaginative names. The list of attempts to fly to Mars (many failed) has some pretty lame efforts when it comes to inspirational names. The first three Russian missions did not qualify for names at all; then came Mars 1, 2, 3 and so on. In more recent times Observer, Surveyor and Pathfinder have given an indication of the purposes of each spacecraft.

Mars Express mission badges below, showing Beagle 2 on the upper face of the spacecraft.

Mars Express, MEx as it is abbreviated (space scientists love abbreviations and acronyms) got its name because it had to happen on a rapid timescale. The idea for Mars Express was first discussed only in 1997 and the spacecraft needed to be ready by 2003 to take advantage of Mars being closer to Earth than it has been for many thousands of years.

At first Mars Express was going to be a mission to orbit the red planet. Beagle 2 got the chance to hitch a ride after a battle worthy of any of the warships which previously bore the name, especially when it came to raising the money.

The Beagle 2 masthead from the project newspaper.

And where is it all going to lead? There can be few people who, when staring at the wonder of the night sky, especially far from the light pollution of land on an ocean voyage, have not asked the question: "Are we alone in the Universe?" FitzRoy and Darwin must have done it. If Beagle 2 finds any trace of life on Mars, it could be the first step to knowing the answer to the ultimate puzzle.

The journey of Beagle 2 could herald discoveries easily on a par with the outcome of the 1830s voyage.

Header expanded to show the destination of the Beagle 2.

The launch

The launch of HMS Beagle*, September 7th 1967. Copyright Brooke Archives.*

HMS *Beagle* (the ninth vessel of this name) was launched on September 7th 1967 at the Lowestoft shipyard of Brooke Marine Ltd by the wife of Rear Admiral G.S. Ritchie, the Hydrographer of the Navy. Her launch made the marine news and, as common with other Royal Navy warships, postcards depicting the ship were produced.

The world's media reported the launch of Beagle 2 from the wastes of Kazahkstan on June 2nd, flawless and precisely on time as planned; coincidently 50 years to the day from the Coronation of Queen Elizabeth II and the announcement of the first ascent of Mount Everest.

In the days leading up to the launch, Beagle 2, now fixed securely to Mars Express, was last seen as the rocket fairing slowly moved into place. After that the rocket was rolled out, moved to the launch pad and raised into the vertical position. Only then was a safety switch activated to connect the Beagle 2 battery into its electronic circuits so that the lander can be sent commands.

The last sight of Beagle 2 as the rocket fairing was positioned over Mars Express.

The rocket rolled out on a slow-moving train to the launch pad.

The launch of Beagle 2 aboard Mars Express, June 2nd 2003. Photograph copyright ESA/Starsem.

The Soyuz rocket soared off the launch pad, a spectacular image against the night sky, a real star before being lost from sight.

Landing on Mars is scheduled for 2.54am GMT on December 25th, Christmas morning; on Mars it will be just after 2pm.

Is there life on Mars?

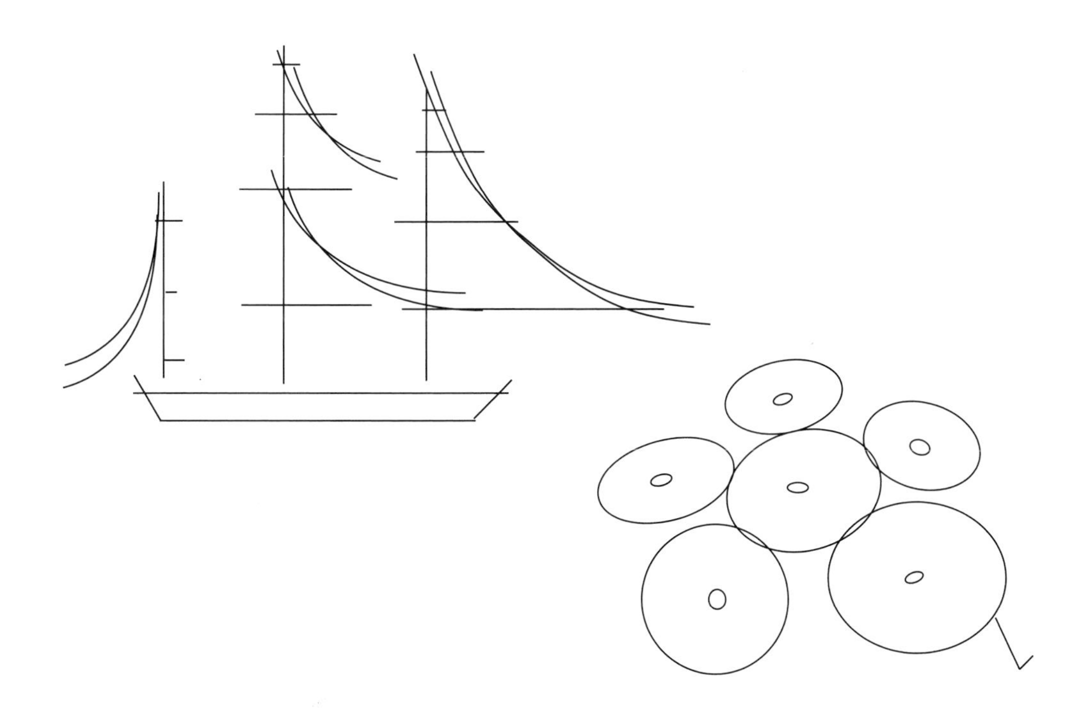

An account of our understanding - and otherwise - of life elsewhere in the solar system and in particular Mars, following progress alongside a history of the various naval vessels named Beagle.

Wandering star

Nicolaus Copernicus (1473–1543) by an unknown artist. Courtesy National Portrait Gallery, London.

Ideas about life in space have oscillated backwards and forwards rather like the erratic passage of Mars across the sky. It was this tendency for retrograde motion which first led to the coining of the term *planet* or *wanderer*. In the time of ancient Egyptian astronomers, it was the deviation of Mars from the conformity of the stars which would have been most apparent.

Although it was acceptable to believe in extraterrestrials more than two thousand years ago, later on the Earth-centred universe of Ptolemy and world domination by the Roman Catholic faith, made it heresy to challenge the teachings of the church. Nicolaus Copernicus, who put the known planets in the correct order around the Sun, only dared to publish his ideas when he was on his deathbed.

Giordano Bruno, a known subversive, who trekked around Europe in the 16th century was not so smart, never being able to stay long in one place once he began to discuss his ideas on Philosophy (science). Recognising a Sun-centred solar system meant thousands or even millions of planets around other stars teeming with life, in 1584 he wrote *Plurality of Worlds*, composed incidentally during a short visit to Britain spent in London and Oxford. When he later returned to his homeland of Italy he was put to death 'without a drop of his blood being spilt'. In other words he was burnt at the stake. It would be tempting to think that Bruno died for his belief in extraterrestrial life but it was probably not so, the Inquisition had a long list of far greater crimes with which to charge him.

Where Giordano Bruno met his end – the Campo di Fiore in Rome showing the gallows in the centre of the horse market; today the area is a bustling market dominated by a statue of Bruno erected in the late 1880s.

The first *Beagle* ship (1766 – ?)

The first ship of the line to be called HMS *Beagle* sailed over the horizon in 1766. She was a gallivat of around seventy tons, capable of being sailed or rowed. The sailors who provided the power acted as a boarding party during naval engagements. The design was peculiar to Eastern parts of the globe in having a triangular sail rather like an Arab dhow. The style also became familiar in Europe, in Holland and northern Germany. The earliest of all the Beagles belonged to the Bombay Marine. Nothing is known about her career in the Navy or what became of her.

The mythical ship that sailed to Mars (William Timlin) is amazingly similar to a gallivat.

Nothing's new in science

In 1837 Charles Darwin opened a new notebook and boldly wrote *Zoonomia* across its title page. In doing so he confirmed the oldest adage in science: nothing's new. The heading under which he would begin to compose his thoughts regarding what he had seen on his travels was the same one

Erasmus Darwin age 38 by Joseph Wright of Derby, his friend and patient. Courtesy of Darwin College, Cambridge and the Darwin Heirlooms Trust.

chosen by his grandfather, Erasmus, to discuss the laws of organic life and what he called 'perpetual transformations', which has to be evolution by natural selection by any other name. Erasmus Darwin's treatise was a medical textbook but this was not all he wrote during his scientific career from 1757 to 1802, for in his poems can be found the beginning of the accusation levelled against Charles by the church that he wanted 'to relieve God of the labour of creation':

> "Nurs'd by warm sunbeams in primeval caves
> Organic life began beneath the waves...
> Hence without parent by spontaneous birth
> Rise the first flecks of animated Earth."

Like Copernicus, Erasmus Darwin's most aetheistic work, *The Temple of Nature* sometimes called *The Origin of Society*, was published after his death. He was roundly condemned on all sides for undermining the resolve of the nation at a time of war.

Unlike his nephew Charles, Erasmus was an inventor and fascinated by mechanical devices. He was very taken with the first attempts at manned flight in hot air and hydrogen balloons. He became an early advocate of space travel, celebrating the Montgolfier brothers, the first balloonists, in verse in another major work *The Botanic Garden:*

> "Rise great Montgolfier! Urge thy venturous flight
> Higher o'er the moon's pale ice reflected light;
> Higher o'er the pearly sun, whose beamy home
> Hangs in the east, gay harbinger of morn;
> Leave the red eye of Mars on rapid wing
> Jove's silver guards, and Saturn's crystal ring
> Leave the fair beams, which, issuing from afar
> Play with new lustres round the Georgian star..."

On a more practical note Erasmus saw the possibility of using balloons to carry manure uphill to save carting across muddy fields.

A plea of insanity

Erasmus Darwin belonged to a generation when, for the first time in history, there was a mass media to comment about the controversial subject of life beyond Earth and in the late eighteenth century space stories began to appear in newspapers.

The first newspaper to mention the subject, in its account of the trial of Dr John Elliot, on July 16th 1787, was *The World*

The World
FASHIONABLE ADVERTISER.

The World *masthead*

and Fashionable Advertiser. The paper reported that the defendant was indicted for wilfully and maliciously shooting one Mary Boydell at half past one of the afternoon of July 9th as she walked past Prince's Square, Soho, on her way to Wimpole Street.

Mr Nicoll, a bookseller of the Strand, her escort, gave his evidence: He was arm in arm with the lady when he heard and felt a shot so close that he thought he had been struck a blow. Turning he perceived the prisoner with two pistols in his right hand. "Are you the villain who fired he cried out, to which the prisoner replied I am the man." With the help of Thomas Griffith, butler, a servant of Mr Brand the surgeon, and a boy in livery, Dr Elliot was apprehended. Indeed he came quietly with them to the magistrates saying "I have my revenge". Only when he found that Miss Boydell had survived his attack did he become violent, "breaking into an incessant vein of acrimonious language against the lady".

At his trial, Elliot betrayed symptoms of a disturbed mind – sometimes staring with "a wild vacancy of countenance, at others lolling with his head on his arms as if asleep". His landlord testified to his madness and his former business partner was prepared to say there was an eccentricity about him, but "so far he hadn't actually poisoned any patients". No wonder his defence lawyer, Mr Sylvester, tried to enter a plea of insanity, calling as his main witness Dr Simmons FRS, the physician at St Luke's Hospital.

Dr Simmons said it was his opinion that the defendant, whom he had known for fifteen years, was mad. He read to the court extracts from a letter written by Elliot the previous January, which he had asked the learned doctor to communicate to The Royal Society. It was the prisoner's theory that the Sun was a body of light, not heat, where day was eternal, and the seasons fruitful. Therefore not only was it capable of being, but was most likely, inhabited. *The World* commented that this did not necessarily make Dr Elliot a mad man, only a speculator. The judge and jury were inclined to agree for they acquitted Elliot. Not on the suggested grounds of diminished responsibility, but because the surgeon who examined the lady found only burn marks on her dress and a bruise on her body. A smoking gun was insufficient evidence. There was no sign of any holes in her whalebone stays where a pistol ball had entered and without a bullet there could be no case to answer. The magistrate was not required to adjudicate on whether life existed beyond Earth.

By the time the *Gentleman's Magazine* carried the same story at the end of July, Dr Elliot had starved himself to death.

Herschel's Mars

William Herschel. Courtesy Royal Astronomical Society, London.

Just a few years after Dr Elliot made the news, William Herschel included a very similar idea about living beings on the Sun in a paper published in *Philosophical Transactions of the Royal Society*, the foremost scientific journal of the day. But Herschel was George III's favourite astronomer, he tried to name the sixth planet Uranus, which he had discovered, 'Georgian Star' after the King and could get away with some indiscretions. Nevertheless, even Herschel was not foolhardy enough to submit his private correspondence and observations concerning trees, forests, and even cities on the Moon for publication.

Despite his wilder fantasies involving the Sun and the Moon, Herschel did seminal work on Mars observing the seasonal evaporation of the polar caps, and their transfer from one hemisphere to the other. He measured the length of the martian day at 24 hours and 39 minutes, very close to what is now the accepted value of 24 hours, 37 minutes and 23 seconds. But perhaps his greatest contribution was to show that the atmosphere was extremely thin by observing the edge of Mars as it passed in front of distant stars.

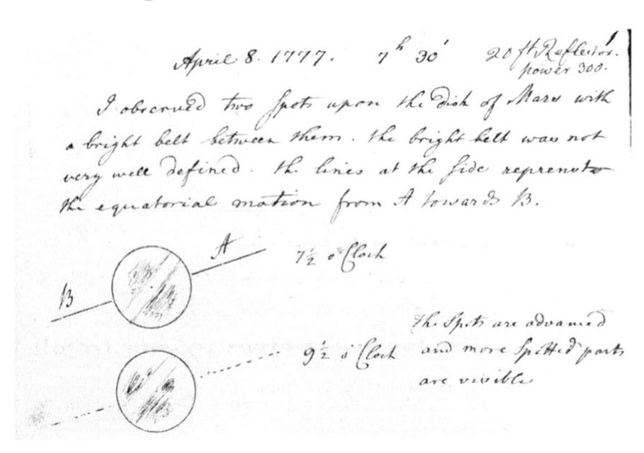

April 8. 1777. 7h 30' 20 ft Reflector. power 300.

I observed two spots upon the disk of Mars with a bright belt between them. the bright belt was not very well defined. the lines at the side represent the equatorial motion from A towards B.

A 7½ o'Clock

B

9½ o'Clock

The spots are advanced and more spotted parts are visible

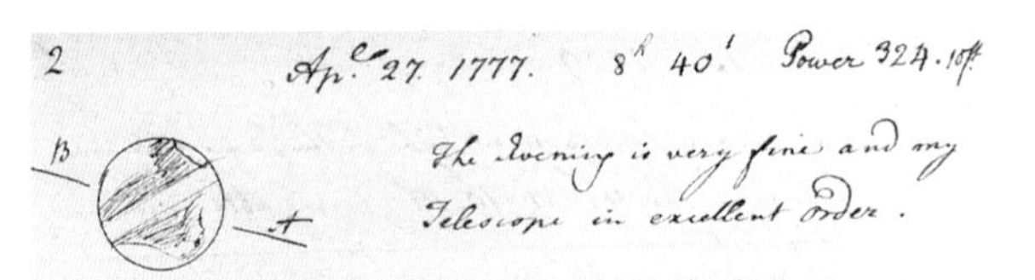

2 Apr. 27. 1777. 8h 40' Power 324. 10ft

B

A

The Evening is very fine and my Telescope in excellent order.

Observations of Mars made and drawn by William Herschel. Courtesy Royal Astronomical Society, London.

Vauxhall Gardens by Thomas Rowlandson c. 1784, depicting the fashionable society of London listening to the orchestra. Edward Topham, on the left in the foreground, looks across to Mary Robinson, the famous actress, and mistress of the Prince Regent, standing next to her. Courtesy British Museum Prints and Drawings, London.

Stones from the sky

It was fitting that the first media story of imagined extraterrestrials appeared in *The World*, since science owes its proprietor Edward Topham a great deal for his persistence on behalf of the real thing – not beings but meteorites.

In the late 1790s people who thought they had observed stones to fall from the sky were not necessarily believed to be mad but certainly they were treated with disdain. They were mostly ignorant peasants who were to be mocked, after all they did not know any better. When a young scientist Ernst Chladni wrote a book on the subject he was greeted with the comment from his Professor "I felt as if I had been struck on the head by one of his stones".

The President of The Royal Society, Sir Joseph Banks, having been informed by the Bishop of Bristol about a supposed meteorite fall in Italy, wrote to Sir William Hamilton (best remembered because his wife ran off with Admiral Nelson) that "the odd bishop is a teller of tall-tales".

Then it happened, the largest meteorite ever to fall in Britain landed in 1795. It chose as its place of arrival Edward Topham's home: The Wold Cottage in Yorkshire. Topham, now in retirement although he was only forty-six years old, did what any self-respecting ex-journalist would have done: he wrote to the papers. And when he was not believed, he kept on writing to the papers and the magazines, and he put his stone on display in a prominent Coffee House in Piccadilly. When they still did not accept his story he used his position as a magistrate to publish sworn

Next to the monument erected by Edward Topham, the author shows the current owners of The Wold Cottage a replica of the meteorite.

The Wold Cottage meteorite, still the largest British meteorite, pictured in a nineteenth century engraving with two other stones, High Possil (fell 1804, Scotland) and Tipperary (fell 1810, Ireland).

testimonies by his workers who had seen the stone fall. Topham was paranoid about the truth; he went on and on and on about how his word as a gentleman should not be doubted.

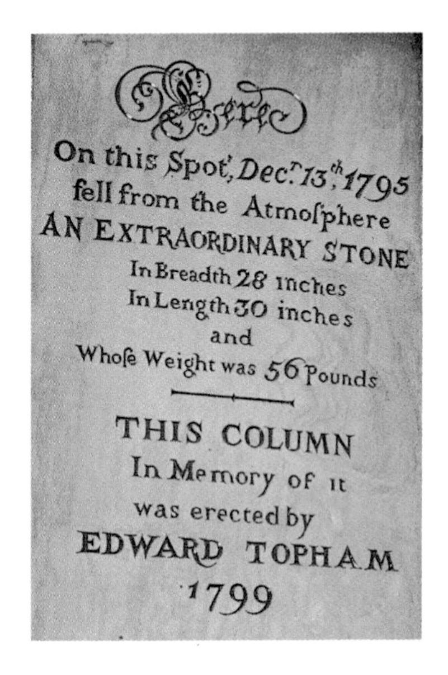

The inscription written by Edward Topham, recently restored with financial support from the Meteoritical Society.

Sir Joseph Banks had little option but to commission a chemical study, carried out by Charles Edward Howard, a brother of the Duke of Norfolk.

Howard gave his verdict to The Royal Society in 1802; it took three meetings but the conclusions were clear. Topham's Wold Cottage meteorite and all the other samples he had been given, no matter where on Earth they had been found, contained metallic iron contaminated with nickel. No known terrestrial rock had these properties, therefore the meteorites must have come from a common source. There was only one place it could be – space!

Europe was convinced but American President, southerner Thomas Jefferson, apparently was not. When a very large meteorite fell at Weston, Connecticut, in 1808, he was reputed to have said "I would rather believe that Yankee Professors lied than that stones could fall from the sky". What is it about meteorites that makes them of interest to American Presidents?

The second *Beagle* ship (1804–1814)

By the time the second ship bearing the name HMS *Beagle* ventured out of the Thames, the British Navy was master of the high seas. The Battle of Trafalgar having been won, albeit at the cost of Admiral Nelson's life, the British fleet were seeking out the remnants of the French Navy wherever it could find them.

One such action involved the latest HMS *Beagle*, a 100 foot brig of 383 tons carrying eighteen guns and a crew of 121 men captained by Commander Francis Newcombe. At Basque Roads a fleet of some sixty British vessels had blockaded the French in port. A plan involving the use of fireships, and vessels packed with explosives, was hatched to break the deadlock. On April 11th 1809, HMS *Beagle*, despite continued bombardment from the defenders, escorted the fire ships across the boom protecting the harbour and in amongst the French fleet. Panic followed, the French cutting their anchor cables in an effort to escape the ensuing conflagration. In the meleé all but two of the French ships ran aground.

Lord Cochrane, who was directing the operation, called upon his superior, Admiral Gambier, to commit half the fleet so that the enemy might be totally destroyed. Despite repeated requests Gambier refused, so Lord Cochrane pressed on with his small force, HMS *Beagle* playing a prominent role. She fought for five hours in water barely deep enough to stay afloat, until Lord Cochrane was ordered to withdraw.

Back in England, by tradition, the Houses of Parliament should have given a vote of thanks to the Commander-in-Chief, Gambier, for a glorious victory (the British losses were limited to eight dead, twenty-four wounded). However Lord Cochrane declared his intention to oppose the passage of the vote through the House of Lords. Gambier was court martialled at his own request and acquitted by a carefully selected and sympathetic court. Eventually he received his congratulatory vote but holds the distinction of being a land-lubber admiral spending only five years at sea in his entire career. Napoleon Bonaparte's comment on his wrecked Navy was "The French Admiral was a fool and the English commander-in-chief no better".

Lord Cochrane was 'framed' in a financial scandal and thereafter became a mercenary spending much of his life helping to establish the republics of South America.

The second HMS *Beagle* continued to harass the French assisting in an attack on the fortress of San Sebastian in August 1813 before being sold in 1814 for £900.

The bit that got away

Chassigny is one of those archetypal villages of France, instantly recognisable by anyone who has driven across the northern part of the country. A long road stretches through its centre, there is a *Mairie*, with a *Tricolore* outside, a *Poste*, a bar and if you see a person as like as not he will be dressed in 'blues'. It has not changed much for 200 years. Here on October 3rd 1815, at 8.30 in the morning, to the accompaniment of what sounded like musket shots – a noise all too fresh in the memories of veterans of the recently ended Napoleonic Wars, an opaque body fell to Earth in a vineyard some distance from the village. The road leading to the spot is now known locally as Le Chemin de Meteorite.

Looking along the lane known locally as Le Chemin de Meteorite.

It was just as well Topham banged the drum for his Wold Cottage stone or the Chassigny meteorite would not have been accepted as an extraterrestrial object. An absolutely key sample for meteoriticists could well have been thrown away as something an idiot peasant tried to pass off as a piece of cosmic curiosity.

The Chassigny meteorite is a piece of rock from Mars although it was one hundred and sixty-odd years before it was recognised as such. As it is a vital bit of Chassigny was discarded. Monsieur Pistollet, the physician from the nearby town of Langres wrote, "having proceeded to the spot in person I collected about sixty small pieces, some of which

Whilst Europe prepares to go to Mars with Mars Express and Beagle 2, few are aware that Mars came to Europe. How many of the good folk of the French village of Chassigny, resting here in the local churchyard, witnessed the meteorite land?

were wet and easily crumbled in the hand". Where did these soft wet bits go? Not to museums in Paris, Vienna or London since the total mass in these collections is well short of the original four kilograms weighed by Pistollet.

If Pistollet's description, variously reported in French, German and English is true, and there is no reason to doubt it, then an important clue in the search for the possibility of life on Mars has been lost. The portion discarded sounds as though it was altered by water. Such portions in other martian rocks contain the evidence of life which prompted the Beagle 2 mission to Mars.

How meteorites are named

From the time of the Wold Cottage meteorite, it became convention to name meteorites from their location of fall, usually from the nearest inhabited place to the actual site. The name, once given, remains unaltered even if it was originally given in error. The same applies if the name of the place of fall subsequently changes over time. In the case of the Antarctic or hot desert samples, a nearby geographical feature followed by a number identifying the year of discovery has become the standard designation.

And places

Orders issued to the next HMS *Beagle* were of much the same vein: "Trifling as it may seem, the love of giving a multiplicity of new and unmeaning names tends to confuse geographical knowledge. The name stamped upon a place by the first discoverer should be held sacred".

The third *Beagle* ship (1820–1845)

HMS Beagle *from* A Naturalist's Voyage Round the World, *published by John Murray 1890. By permission of the Syndics of Cambridge University Library.*

The third *Beagle*, Darwin's ship, was launched at Woolwich on May 11th 1820. She was a Cherokee class brig of 235 tons, ninety feet in length and capable of carrying seventy-three men.

The Cherokee-class vessels earned the uncomplimentary name of 'coffin brigs'. They had a very shallow draft hull which gave them easy access to coastal waters. The design was not so good however for sailing on the high seas which might account for Darwin's continual sea-sickness and the rather unkind reference to the undertaking profession. It was ideal for the Cherokee's primary purpose, swift and manoeuvrable for engagement with French pirates and American buccaneers who were not yet ready to concede that the Napoleonic wars were over.

But indeed they were and *Pax Britannica* had descended with the result that the third *Beagle* was placed 'in ordinary', what would now be called mothballed, for five years before being given a new lease of life as a survey ship.

It was for this activity that *Beagle* set sail in 1831, although the orders acknowledged that the voyage was also a scientific undertaking. Darwin was on board as a passenger "to relieve the proverbial loneliness of the Captain" as Robert FitzRoy put it. On *Beagle*'s first voyage of exploration the original commander, Pringle Stokes, had shot himself whilst in a fit of depression. His aim must have been bad since he took ten days to die.

Robert FitzRoy, Captain of the third Beagle. *"Presented to Greenwich Hospital by the surviving officers of HMS* Beagle *as a memorial of their esteem and regard for their distinguished Captain." Courtesy of Greenwich Hospital Trust and Shrivenham Military College.*

Knowing he was prone to melancholy moods, the twenty-six year old FitzRoy, who replaced Stokes, took Darwin as a companion, someone of his own standing, for conversation as the Captain was expected to remain aloof from the crew to maintain discipline. So the young man of twenty-two years, who had failed at becoming a medical doctor and who had no liking for the Church, began his third career aboard His Majesty's Ship *Beagle*.

A lunatic hoax

In the 1830s, the public were obviously ripe for the discovery of extraterrestrial life. The great lunar hoax, for that is how it became known, was the work of an expatriate British journalist who would later try to claim he was writing satire. It began on the morning of Tuesday August 25th 1835, when

Charles Darwin, 1840, by George Richmond. Courtesy G.P. Darwin on behalf of Darwin Heirlooms Trust, copyright English Heritage.

HMS *Beagle* lay off the coast of Chile poised to cross the Pacific on her way to the Galapagos. An insignificant headline on the front page of the *New York Sun* declared the one penny daily paper would publish "Great Astronomical Discoveries" lately made by Sir John Herschel (son of William) at the Cape of Good Hope with his new twenty-four foot diameter reflector capable of magnifying 42000 times. A preview of what had been observed by Sir John over the period of the 10th to the 14th of January followed to show that he had "affirmatorily settled the question whether this satellite [the moon] be inhabited, and by what order of beings".

A papier-mâché snuff box, English circa 1835, illustrating the astronomer Sir John Herschel and the lunar hoax. Courtesy The Royal Astronomical Society and the Museum of History of Science, Oxford.

Fascinated New Yorkers boosted the paper's circulation to nineteen thousand copies daily during the revelations. The first instalment contained information about geological and botanical discoveries. Sir John's fictitious but seemingly brilliant astronomical technician, Dr Andrew Grant, was the supposed source of the leak to the paper using material in preparation for publication by the *Edinburgh Scientific Journal*. On the 27th, a biped beaver was revealed. This creature was acquainted with fire, carried its young in its arms and lived in huts better than those of human savages. Grant was quoted as saying "then our magnifiers blest our panting hopes with specimens of conscious existence". Little wonder hundreds were queuing in the streets on August 28th for the first editions. They were not disappointed by the four foot high humanoids with copper-coloured hair and transparent wings, given the name *Vespertilio Homo* or Manbat. These were undoubtedly intelligent beings, they could be seen having conversations.

The delighted proprietor of the newspaper, Benjamin Day, who was not party to the deception, by now was offering its articles as a pamphlet, sixty thousand copies were to be made available to an eager clientele. Lithographs were also promised. The discoveries however were getting more and more extravagant, magnificent temples dedicated to devotion were reported and observations made of a higher order of more beautiful beings. On the 31st the paper hinted that studies were to move on to planets. Who knows what was in the pipeline for Mars amongst the forty pages of illustrative material that the *Sun* declined to publish in the interest of keeping the price of the paper in bounds.

Strangely other newspapers in their haste to carry the story, instead of going back to the original source of the material, were taking for granted the *Sun*'s text. They were praising the work, the *New Yorker* said "these discoveries create a new era of astronomy and science in general". The *New York Times* congratulated the author for the most accurate and extensive knowledge of astronomy, and, despite gobbledegook about hydro-oxygen light, reported news of a £10,000 grant from the King. It claimed the story was being verified by civil, military and religious authorities.

Richard Adams Locke.

It was only when a reporter named Finn, from the *Journal of Commerce*, called at the *Sun* to seek official permission to reproduce material that the deception came to light: he was met by its perpetrator Richard Adams Locke who admitted making the whole thing up. Finally the *New York Herald* announced the fraudster's identity.

Sir John Herschel was by no means displeased about his name being taken in vain possibly in view of his father's earlier lunar life theories. Sir John's wife wrote to their comet-spotting Aunt Caroline, "It is only a great pity that it is not true, but if grandsons stride on as their grandfathers have done, as wonderful things may yet be accomplished".

Sir John Herschel

In 1836, just after the great lunar hoax, Darwin and FitzRoy visited Herschel at the Cape of Good Hope at the end of their journey round the world. Darwin had been inspired by Sir John Herschel, especially his book *A Preliminary Discourse on the Study of Natural Philosophy* and was eager for the encounter. He had read Sir John's work at Cambridge and claimed it "stirred me up with a burning zeal to add even the most humble contribution to the noble structure of Natural

A day or two after his encounter with Herschel, Darwin visited Napoleon's tomb on St Helena. A drawing by Syms Covington, Darwin's manservant on HMS Beagle*. Courtesy Mitchell Library, State Library of New South Wales.*

Science". Afterwards it appears he felt his hero worship misplaced; in his diary of the *Beagle*'s voyage, Darwin related a description of Herschel's demeanour during the visit, "he always came into the room as though he knew his hands were dirty [Herschel was a keen gardener] and that he knew that his wife knew that they were dirty". Unfortunately Darwin probably would not have known that Herschel had been misrepresented by the American newspapers or he could have asked the astronomer's views about extraterrestrial life.

Maps of Mars

Whilst New York's citizens scrambled for their newspapers, and Darwin blissfully indulged in polite conversation as a house guest, some real work concerning Mars was going on in the Tiergarten of Berlin. In the private observatory of Wilhelm Beer, Johann von Madler was compiling the first authentic map of Mars based on observations made with a state-of-the-art three and three quarters-inch telescope. With it the pair were responsible for refining the period of rotation of Mars and set the undisputed zero point of martian longitude, fully fifty years before it was agreed that Earth's meridian would pass through Greenwich.

The fourth *Beagle* ship (1854–1862)

Darwin's *Beagle* left the service in 1845 to be replaced by the first mechanically driven version of the ship to carry the

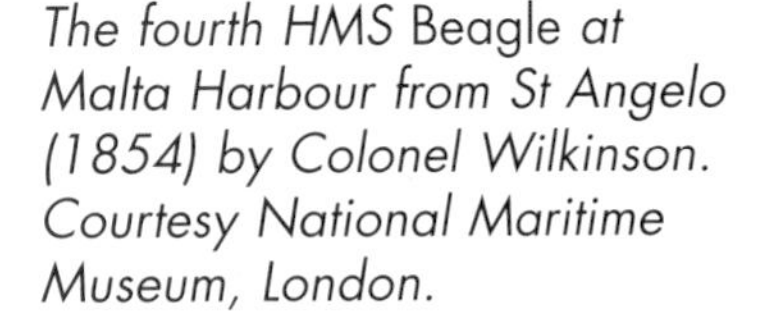

The fourth HMS Beagle *at Malta Harbour from St Angelo (1854) by Colonel Wilkinson. Courtesy National Maritime Museum, London.*

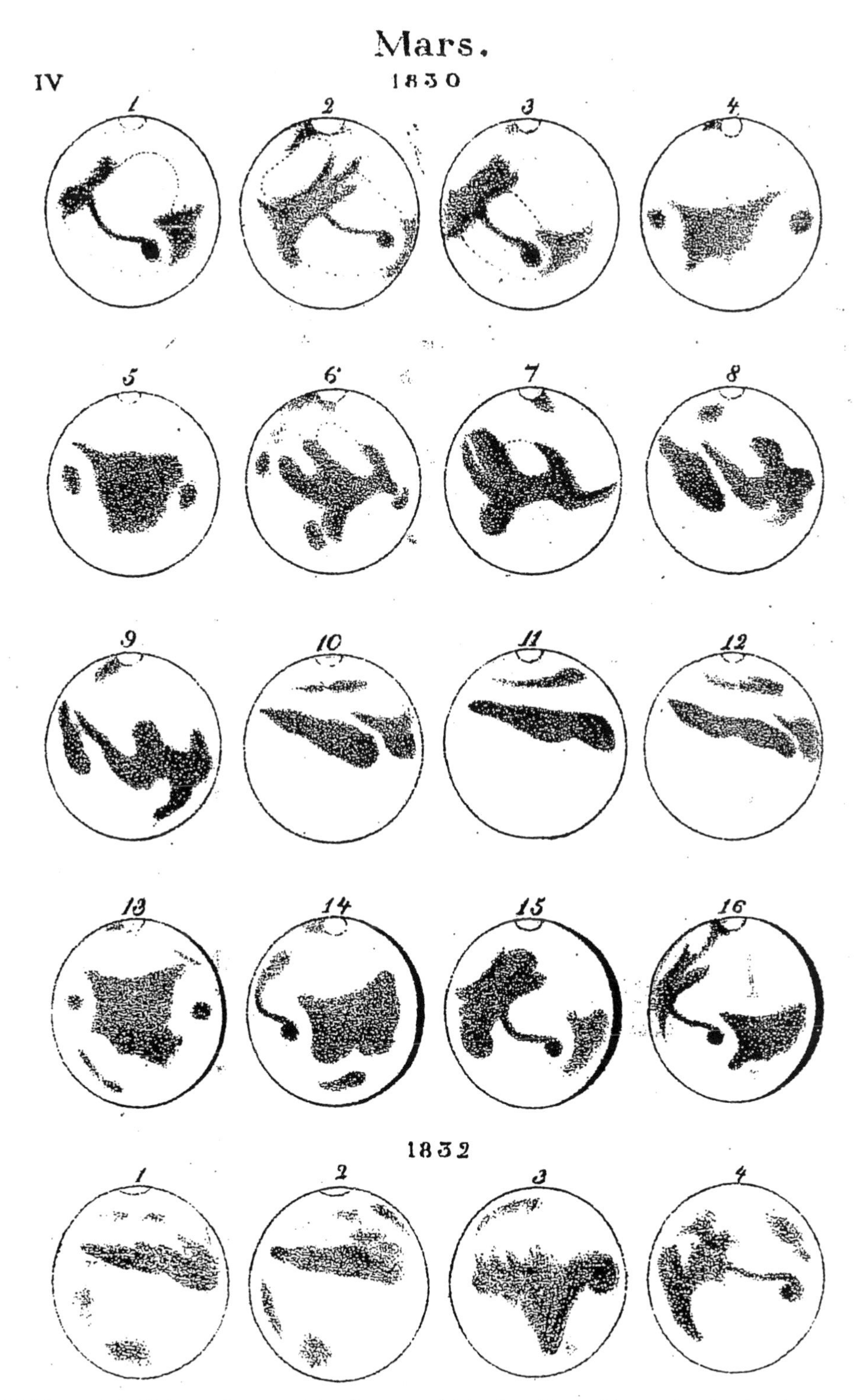

Mars in 1830 and 1832 as observed by Beer and Madler. From Physische Beobachtungen des Mars bei seiner Opposition im Jahre 1841, published in Astronomische Nachrichten, 1842. Courtesy Royal Astronomical Society, London.

name in 1854. The fourth HMS *Beagle* was 477 tons, 160 horsepower, 160 feet long and carried a crew of sixty-five men at speeds of up to eleven knots.

Built at Blackwall on the Thames, the new Beagle was almost immediately dispatched to the latest theatre of war, the Crimean peninsula in the Black Sea. There she more than distinguished herself. On October 25th 1855, the day after the infamous, but glorious, charge of the Light Brigade, an HMS *Beagle* shore party, commanded by Acting-mate William Nathan Hewett, were involved in their own heroics. When the Russians on shore launched a determined attack to repel the invaders, Hewett was ordered to withdraw. But instead of spiking his gun, the young non-commissioned officer used it to blow away part of the battery wall to give a line of fire on the advancing Russians, "obliging them to withdraw". Hewett was promoted Lieutenant that day and later awarded the Victoria Cross, one of the first recipients of a medal struck from metal obtained by melting the guns captured at Sebastopol.

HMS *Beagle*'s acts of courage during the Crimean War did not cease there. On October 11th she rescued a store ship laden with hay for the use of the Army right from under the noses of the Russians. The next year in July one of *Beagle*'s crew, now commanded by Hewett, gained a second VC. This time the recipient was Able Seaman Joseph Trewavas who, with the enemy firing at him from eighty yards away, calmly sawed through the hawsers of a bridge which needed to be destroyed to cut the Russian supply lines. Trewavas was slightly wounded, one of *Beagle*'s absurdly low casualty list of two.

Victoria Cross awarded to William Nathan Wrighte Hewett. Courtesy National Maritime Museum, London.

William Nathan Hewett

Cornishman Joseph Trewavas

HMS Beagle *with HMS* Algiers *and HMS* Samson*, in the Black Sea, November 14th 1854, artist unknown. Courtesy National Maritime Museum, London.*

They were in action again a few weeks later having a hand in the bombardment of Fort Kumbrun from the sea. This occasion was the first time a sea battle employed only steam driven vessels.

Surplus to requirements, at the end of the Crimean hostilities, the fourth *Beagle* was sold for £5500 and departed for the Far East. An indication of a visit to Hong Kong is reportedly in the Happy Valley cemetery – a tombstone erected by sailors from HMS *Beagle* in memory of a colleague. She ended up as a Japanese punishment ship "a place of chastisement" renamed the *Kenko Kan*, at Yokosuku. Here she was mistaken for Darwin's *Beagle* by the Reverend V. Marshall Law, an error perpetuated in the science journal *Nature*. She was scrapped in 1889.

Origin of Species – only the strong survive

The idea of natural selection never came to Darwin in a flash of inspiration, like a bolt from the blue. His ideas were about twenty years in gestation, formulated during hours of wandering around the grounds of his home in Kent, Down House, on what he called the sand path. For the most part, after his return from South America, Darwin enjoyed a gentleman's life of leisure with old friends to visit him as his family grew steadily to ten children. Cataloguing the entire collection of barnacles brought back by HMS *Beagle* was a painstaking task whilst his great theory took shape.

But he should have been stirred into action by a dozen pages published in 1855 in the unpretentious little journal *Annals*

Antarctic dry valley, a place where the rocks contain cryptoendoliths, an example of organisms able to survive under extreme conditions. Courtesy of the late Dr David Wynn-Willliams, British Antarctic Survey.

and Magazine of Natural History. The author, Alfred Russel Wallace, a professional specimen collector, made his living, and paid his exploring expenses, by selling material to gentlemen philosophers, including Darwin. Wallace's thesis in the *Annals* was that "every species has come into existence coincident both in space and time with a pre-existing closely allied species". Surprisingly Darwin did not feel threatened by Wallace. Perhaps his attention was distracted by the words "come into existence" which suggested creation rather than evolution. Instead of expediting his own publication, Darwin wrote Wallace an encouraging letter, told him he was two years away from completing his treatise, and enlisted the collector's aid in obtaining more bird specimens from Borneo.

On June 18th 1858 Charles received Wallace's response which had been dashed off in February from Ternate in the Moluccan Islands. Whereas it had taken Darwin half a lifetime, Wallace had recognised how natural selection worked in two hours of lucidity during a malaria-induced fever. Darwin stared at the pages he had been sent by Wallace with a request to submit them for publication and realised he was reading the abstract of his own life's work. For years he had been tormented by the damage which natural selection – man descended from Apes – might do to his reputation. Now he was devastated, pre-empted by a professional collector, a man who had no such qualms.

A rescue operation had to be mounted, although Darwin said he would rather burn his book than behave less than honourably. His friends Lyell and Hooker stage-managed a joint reading of the Darwin-Wallace theory on June 30th at the Linnean Society. Darwin did not attend the proceedings, instead an 1844 essay and an 1857 letter to botanist A.S. Gray were read out along with Wallace's letter to Darwin dated February 1858. For all Darwin's fears, the meeting ended with the President of the Society lamenting that there had been no outstanding revelations in the year just passed.

It took a further six months for the mail to go back and forward to Wallace; fortunately he was not upset by what had been arranged. Darwin would later remember the debt he owed Wallace, acquiring for him a government pension.

The great debate

The great debate over the ideas put forward simultaneously by Darwin and Wallace did not occur until two years later in Oxford at the 1860 Meeting of the British Association for the Advancement of Science (BAAS). It was chaired by Henslow,

Charles Darwin circa 1855, shortly before the publication of On the Origin of Species, by Maull and Polyblank. Courtesy National Portrait Gallery, London.

the man who suggested Darwin for the role of naturalist on the *Beagle*, and attended by FitzRoy, her Commander. The main protagonist for the evolutionists was Thomas Henry Huxley (another man who had served as a ship's naturalist) and, representing the God-fearing Creationists, Bishop "Soapy" Samuel Wilberforce who earned his nickname from his habit of wringing his hands whilst preaching. Darwin was "too ill" to make the journey from Kent.

A television dramatisation of the affair once depicted FitzRoy as a religious zealot waving a bible and vehemently denouncing his former companion. The contemporary newspaper report of the events makes no mention of an angry outburst from, the now Admiral, FitzRoy. What does survive is a letter sent to Darwin at the time of the publication of *On the Origin of Species*. It states quite simply "My dear friend ... I, at least, cannot find anything ennobling in the thought of being descended from even the most ancient ape".

If there was any anger intended in the missive Darwin never took offence. He was one of the friends of FitzRoy who settled the Admiral's debts on his death a few years later in 1865.

Soapy Sam was less than kind to Darwin, in his review of *On the Origin of Species* he reminded the World that Charles was restating his grandfather's ideas. At first Darwin disowned the theories expounded by Erasmus in *The Temple of Nature*, he was later to regret his actions and tried to make amends in a biography of his forbear.

A canal on Mars

Angelo Secchi. Thanks to Richard McKim.

It was during the era of the fourth HMS *Beagle* (although Darwin's *Beagle* was still enjoying retirement in the Essex Marshes) that the infamous word canal was first used in respect of Mars. Italian Jesuit Angelo Secchi made detailed drawings of Mars in 1858, choosing to call the largest and longest observed feature on Mars, the dark region of Syrtis Major, the Atlantic Canal. In 1860 another astronomer refuted the interpretation that the feature was water and suggested it might be vegetation. The pace of the investigations of life on Mars was hotting up.

Spontaneous generation

Darwin's theories caused religion a problem. How did life on Earth begin if God was not responsible? As early as 1821, when the third HMS *Beagle* was 'in ordinary', the Count de Montlivault supposed that life did not first appear on Earth at all but arose having been transferred from the Moon via the explosive eruption of lunar volcanoes.

Although the development of maggots in meat from flies eggs and sexual reproduction by the fertilisation of ovae by sperm

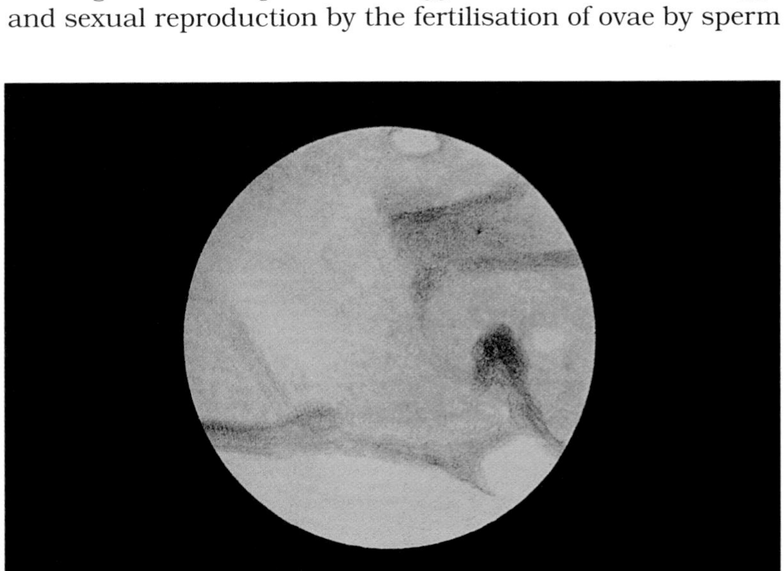

Mars showing Syrtis Major observed and drawn by Patrick Moore in 1969. With thanks to Sir Patrick Moore for permission to reproduce.

were seventeenth and eighteenth century discoveries respectively, until the 1850s it had been believed that microscopic organisms, mould etc, could generate spontaneously from the inorganic world. The carefully conducted experiments of the famous Frenchman, Louis Pasteur, disproved this fallacy.

Louis Pasteur (1822–1895).

Post-Pasteur, it was becoming increasingly obvious that meteorites, particularly a type called carbonaceous chondrites, had ten percent water and were chock full of organic matter. Thus it was only a short step for Hermann Richter to suggest in 1865 that they provided raw material for life on Earth.

Meteorite types

Meteorites occur in several forms: stones, irons and a mixture of the two constituents. By far the greatest proportion are stones and of these the most common type are referred to as ordinary chondrites. The name chondrites derived from the occurrence of tiny, often spherical, inclusions called chondrules. Chondrites show no signs of ever having been completely melted and are thought to represent the original material of the solar system, some four and a half billion years old. Some chondrites have elemental carbon and carbon compounds present, these have become known as carbonaceous chondrites with those richest in carbon being type 1 or type I (after a particular specimen from Ivuna). Some stony meteorites are not chondrites but have been formed by volcanic processes – martian meteorites are included in this group.

A postcard from the Musée Pasteur in Paris, showing historical artifacts in Pasteur's laboratory.

An Indian connection

The Shergotty stone which fell near Sherghati in India. Thanks to Professor N. Bhandari, Ahmedabad.

In 1865, far away on the other side of the world, another piece in the jigsaw which would eventually lead to Beagle 2 became available. On the morning of August 25th, a new meteorite type fell at Shergotty in the province of Bengal, India. At first one government official doubted it was a meteorite. It did not contain metal, the key constituent recognised by Charles Howard, or chondrules, which characterise the common chondrites. But fortunately the deputy opium agent Mr Pepper was more confident. He was right; Shergotty was a meteorite but it was different from other meteorites, having being formed volcanically. For over a hundred years the real significance of Shergotty was not recognised.

When worlds collide

Lord Kelvin (Sir William Thomson) photographed in 1902; he stuck to his ideas concerning meteorites bringing life to Earth until the end of his career.

As Sir William Thomson, later to be Lord Kelvin, stood up to make his Presidential address to the British Association of the Advancement of Science in Edinburgh on August 3rd 1871, it is doubtful whether anyone expected anything controversial. Speaking just a little way away from where Charles Darwin had begun his abortive medical career, and worried about how Darwinian theories of evolution affected his own specialist research – the age of the Earth – Thomson decided to help out, by suggesting life had arrived on our planet *via* meteorites.

Bishop Ussher's religious interpretation that the world began during the evening of Saturday, October 22nd, 4004 BC had already been superseded by scientific estimates, but Sir William still had a problem. He could not see how there was sufficient time for men to develop from microbes in a hundred million years, which was his estimate of the formation age of the planet, the time he calculated it would take a white-hot ball of rock to cool down.

Thomson's public pronouncement was that life arrived on Earth "through moss grown fragments from the ruins of another world". He added "it seems wild and visionary, all I maintain is that it is not unscientific". He predicted martian meteorites at a time when a key one had just fallen to Earth unrecognised.

Of course what the President of the BAAS had said made the next morning's *Times* newspaper. More than that he was twice lampooned in the satirical magazine *Punch*. Even though he held deeply religious convictions, Thomson became unpopular with both the evolutionists and the creationists by shifting the onus for the beginning of life to another world.

The comments he drew however were not all adverse. His long-term friend, Hermann von Helmholtz, a German scientist of equal stature, leapt to Sir William's rescue. Not only did von Helmholtz say he was inclined to agree but claimed, in time-honoured scientific fashion, "I had mentioned the same view as a possible mode of explaining the transmission of organisms through space, even a little before in a lecture". The text of his paper, given at Heidelberg and Cologne in 1871, was not published until 1876.

Von Helmholtz had different reasons for wanting to move life around the Universe on planetary fragments. He had made estimates of how long the Sun could carry on burning which, in the absence of any knowledge of nuclear reactions, were far too short for life to continue to thrive for long. By having organisms transportable by meteorites, Thomson and von Helmhotlz were providing a scientific version of religion's life everlasting.

NO CONJUROR'S CONJECTURE.

COULD a Meteoric Stone,
Pray, SIR WILLIAM THOMSON,
Fall, with lichen overgrown?
Say, SIR WILLIAM THOMSON.

From its orbit having shot,
Would it, coming down red-hot,
Have all life burnt off it not?
Eh, SIR WILLIAM THOMSON?

Not? Then showers of fish and frogs
Too, SIR WILLIAM THOMSON
Fall; it might rain cats and dogs.
Pooh, SIR WILLIAM THOMSON!

That they do come down we're told.
As for aërolite with mould,
That's at least too hot to hold
True, SIR WILLIAM THOMSON!

First published in Punch, August 19th 1871, Sir William's ideas of life coming to Earth via a meteorite were lampooned in verse.

Pride of France

Sir William Thomson's speech at Edinburgh on the state of scientific research drew attention to Darwin not only but also to the work of Pasteur.

Following Thomson's flattering comments on the virtues of his work, Pasteur returned the compliment and decided to study meteorites. He acquired a suite of material from Stanislas Meunier, the curator of the Natural History Museum in Paris, to look for viable organisms. Since none were found, there is no publication describing the effort, only statements made years later by other authors, including Meunier, saying that it had been done.

It is not known whether the samples studied included Chassigny, France's martian meteorite. Pasteur certainly employed his techniques with the country's most recently acquired specimen, Orgueil. The most famous type I carbonaceous chondrite, noted for its abundance of organic matter and water, had fallen in 1864 in a village whose name translates as 'Pride'.

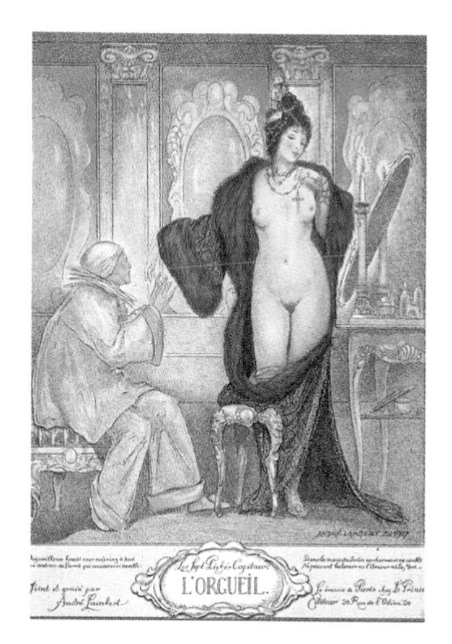

A French picture postcard from a series illustrating the seven deadly sins – according to the traditional view 'Pride' or 'L'Orgueil' is first on the list.

The garden where some of the meteorite stones were recovered near Orgueil. Photograph courtesy of the late Professor Bart Nagy.

The fifth *Beagle* ship (1872–1883)

The fifth ship of the name HMS *Beagle* was built and launched in Sydney, Australia. She was a schooner of only 120 tons and 80 feet in length. Delivered in 1872 and active until 1883, she was one of four such vessels commissioned in response to public outcry following the murder, by natives in the Swallow Islands, of the South Seas missionary Bishop Patteson. A seaman from *Beagle*, Alfred Hodge, wrote a poem in remembrance of shipmates from a vessel called the

HMS Beagle*, painted by F.M. Boyer in 1878. Courtesy National Maritime Museum, London.*

Sandfly killed in similar circumstances. The fifth *Beagle* was almost the smallest of her ilk but undoubtedly she created the biggest fuss.

Five years after the new HMS *Beagle* was commissioned, an Englishman named Easterbrook was murdered on the island of Tanna; the missionary Mr Neilson reported it to Her Majesty's Consul at Noumen. The Consul spoke to the Naval Commander, Captain Hoskyns, who despatched the *Beagle* under Lieutenant Crawford Caffin with explicit instructions – "if after enquiry he should be fully convinced that it was not the misconduct of Easterbrook that led to his being murdered, he was to cause the murderer to be executed according to the judicial forms in the most public manner possible".

On January 24th 1878 Sir Charles Dilke rose from his seat in the House of Commons and asked Mr W.H. Smith, First Lord of the Admiralty, "is it true that a native was hanged from the yard arm of HMS *Beagle*?" He went on to question whether the commanders of HM ships had the competence and the right to pronounce and carry out capital sentences. He was answered, "yes it was" and "yes they do". As politicians are sometimes wont to do, the First Lord added "a short time ago (i.e. under the previous government) we would probably have shelled the village and reduced it to rubble with great loss of life". Four days later the Attorney General was summoned and asked to give his opinion. He affirmed that in places where there was no law prevailing, an attack

The man who raised the Beagle affair, Sir Charles Dilke. He was tipped for high office until cited in a messy divorce.

Mr W.H. Smith, first Lord of the Admiralty (and the son in W.H. Smith and Son, Booksellers).

on a British subject could be considered the same as an act of war and could be dealt with by the Officer in charge who had all the powers he needed.

It seems the Opposition were intent on making an issue of this matter to the point of attempting to charge everybody connected to the trial with a judicial murder. Whether they had some prior information that everything in this case was not as it seemed, or whether someone's instinct said keep digging is unknown, but question after question concerning the affair on the fifth HMS *Beagle* was asked over the next eight months.

Piece by piece it emerged, the complaining Missionary had been part of proceedings although he denied it, admitting only to being the interpreter at the trial. Then it transpired that Caffin had not hanged the murderer at all, just an accomplice. It was alleged that nine natives were hanged although the Navy said these persons were summarily dealt with by the tribe in an effort to identify the culprit. In defence of HMS *Beagle*'s actions, the Government stated ten white men had been murdered in the last nine years without anyone being brought to justice.

Sir John Gorst, a future Government Attorney General, had the last word.

In a full debate on August 6th, Sir John Gorst putting the motion said he wished "to establish some Parliamentary check on a system which turned naval officers into judges and Her Majesty's ships into perambulating gallows". He required that "orders be issued to officers commanding Her Majesty's ships to define and regulate their authority to put persons to death".

The full story (if that is possible without the unfortunate native's version of events) then emerged. When the Chief of the Islands refused to give up the murderer, Lieutenant Caffin took as hostages all those natives currently on board the *Beagle*, one of whom happened to be the local headman. This simple expedient resulted in the information that a native called Nakapok had stolen some coconuts from Easterbrook and then rather cheekily tried to sell them back. The pair argued for days about the coconuts, then Easterbrook, to get his revenge, 'took liberties' with a woman belonging to Nakapok called Yasagu (she, on questioning, admitted she was a willing party). Nakapok retaliated by inducing a man called Yumanga to shoot Easterbrook; the assassin took with him his younger brother, Nokwai, carrying a club. After the deed, Nakapok and Yumanga had fled leaving Nokwai, who admitted his presence at the scene knowing that a murder was to be committed. The Attorney General stated that under British law all three were equally guilty.

After a long argument concerning the legal niceties, the House was reminded that the powers that were being questioned allowed British officers "to protect the lives and property of those who had no power to protect themselves against persons who really were outlaws even if some of them bore the name of Englishmen". On this Gorst withdrew his motion.

On guard

There is no doubt as far as the Navy was concerned that Caffin was within his rights. The orders issued to Robert FitzRoy contained the following passage:

"The narrative of every voyage in the Pacific abounds with proof of the necessity of being unremittingly on guard against the petty treacheries or more daring attacks of the natives. It should be recollected that they are no longer the timid and unarmed creatures of former times but that many of them now possess firearms and ammunition and are skilful in the use of them. Temper and vigilance will be the best preservatives against trivial offences and misunderstandings which too often end in fatal quarrels; and true firmness will abandon objects of small importance where perseverance must entail the necessity of violence; for it would be a subject of deep regret that an expedition devoted to the noblest purpose, the acquisition of knowledge, should be stained by a simple act of hostility."

The third HMS Beagle *occasionally met with hostile natives, most famously on the coast of Australia when Messrs FitzMaurice and King danced for their lives during the third voyage (from* Discoveries in Australia *by John Lort Stokes, 1846, by permission of the Syndics of Cambridge University Library).*

The subject has not been debated since so it must be assumed it is still legal for lawless persons to be executed by the Commander of a vessel having the name *Beagle.*

By 1879, the Navy had returned to a policy of bombardment to exact retribution; it must have seemed altogether less trouble.

A very fertile mind

If 1870 was the time for organisms in meteorites, 1880 saw the turn of the fossil hunters. The discoveries in question belonged to Dr Otto Hahn, not the atomic physicist who would try to date meteorites using the newly discovered radioactivity in 1912, but a lawyer who had studied geology whilst at the University of Tubingen.

In 1878, Carl Gumbel in Munich first scrutinised polished thin sections of carbonaceous chondrites, like the French meteorite Orgueil, for signs of organic debris. The amounts of carbon in these specimens were so large that chemists investigating them had equated the organic matter to peat or even coal which they knew to have originated from plant life. Gumbel was unsuccessful in his search for fossils but prophesised that someone else might be luckier with more 'fertile' material. He had to wait little more than a year before Hahn obliged with descriptions of algae, sponges, corals and crinoids, not from carbonaceous chrondrites but from ordinary ones, most notably a Russian sample called Knyahinya. The *pièce de résistance* of Hahn's fertile mind was that he interpreted the Widmanstätten pattern (a distinctly non-biological, effect produced from interlocking crystal growth), in the iron meteorite Toluca, as a fossil fern.

A typical Widmanstätten pattern in an iron meteorite, a distinctly non-biological effect due to interlocking crystal growth. With thanks to Dept of Earth Sciences, Open University.

Initially Hahn was strongly supported by the press: *Popular Science Monthly* stressed "the transcendent importance of this new and great discovery".

It was said that Hahn journeyed to see Darwin at Down House, laid his specimens before the aged founding father of evolution causing him to leap from his chair exclaiming "Almighty God! What wonderful discovery, now life reaches down". Down House has no record of the visit, but two letters from Hahn exist, one saying he was sending a copy of his book to Darwin. Another letter, from Thomas Bonney, Professor of Geology at the University of London, opposed the idea of meteorite fossils.

The sensation was short-lived. Soon the scientific establishment were pointing out to Hahn the error of his ways; he had in most cases mistaken the ubiquitous spherical chondrules, made of silicates, for residual organic remains. *Popular Science Monthly*, which had originally been so enthusiastic, now poured scorn on the idea, including a quotation from a noted expert that Hahn's "imagination has run wild with him". They made no comment about the journal's imagination nor did they seek the opinion of Kaiser Wilhelm, to whom the discoveries had been dedicated.

Writing much later in 1904, British meteorite expert Sir Lazarus Fletcher commented that it must have been a practical joke.

Maps galore

The late 1870s were a great time for martian map-makers, in particular Giovanni Schiaparelli.

Schiaparelli was Director of the Observatory in Milan from 1862; his career as a watcher of meteors and comets was exemplary, earning him a gold medal from the British Royal Astronomical Society (RAS). He first observed the features on Mars which he was to refer to as "canali" (channels) with an eight-inch telescope between September 1877 and March 1878. It was one of his earliest opponents, artist, art teacher to Queen Victoria and sometime Mars map-maker, Nathaniel

Map of Mars by Schiaparelli showing the canal network. Courtesy Royal Astronomical Society, London.

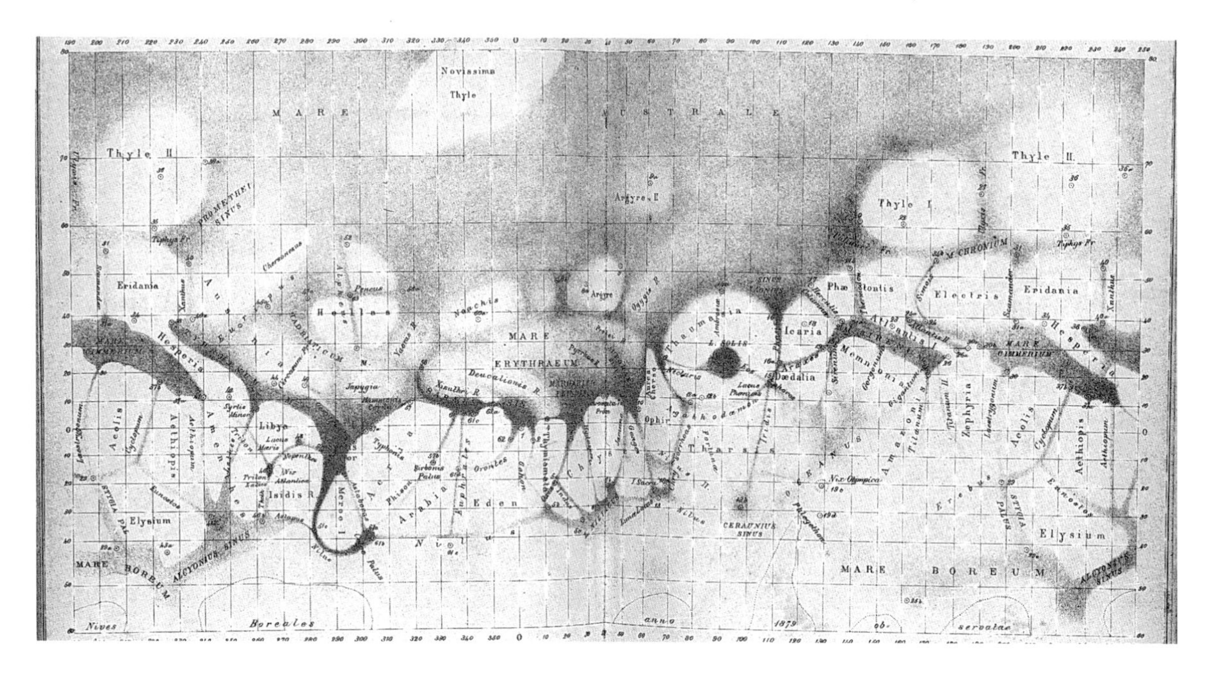

Richard A. Proctor, astronomer, by Sir Leslie Ward, the caricaturist known as Spy, published in Vanity Fair March 3rd 1883. Courtesy National Portrait Gallery, London.

Green, who was the perpetrator of the mistranslation to *canals* that began all the fuss.

News of Schiaparelli's canals broke in the English press in the form of correspondence by T.W. Webb from his vicarage in Hereford to the Editor of *The Times* on April 10th 1882. A few days later *Scientific American* reported a version acquired from the *Daily Telegraph*. As is always the case, the initial letter prompted a rash of 'who saw it first' follow-ups. British astronomer, Richard A. Proctor, informed *The Times* on April 13th 1882, the honour should go to W.R. Dawes for his 1862 observations.

CANALS ON THE RED PLANET MARS.

TO THE EDITOR OF THE TIMES.

Sir, – I possess 30 or 40 views of Mars presented to me 16 years ago by the Rev. Mr. Dawes ("eagle-eyed Dawes", as he was aptly termed), in which, though he used but an 8in. telescope, some of the long, narrow passages mentioned by Mr. Webb are shown. I mention this because it may serve to corroborate what otherwise might seem improbable – the circumstances that Signor Schiaparelli should have seen with his comparatively small telescope what has escaped the attention of observers using such instruments as the Herschelian reflectors, the 3ft. reflector made by Mr. Common, and the magnificent 26in. refractor of Washington. Albeit until observers with such instruments as these have distinctly seen what Signor Schiaparelli has mapped we must not too hastily assume that these are real features of Mars. Mr. Nathaniel Green, whose fine lithographs of Mars adorn a recent volume of the "Memoirs of the Astronomical Society," considers that these narrow passages are due to an optical illusion (which he has himself experienced.)

Should it be proved that the net-work of dark streaks has a real existance, we should by no means be forced to believe that Mars is a planet unlike our earth; but we might perhaps infer that engineering works on a much greater scale than any which exist on our globe have been carried on upon the surface of Mars. The smaller force of Martian gravity would suggest that such works could be much more easily conducted on Mars than on the earth, as I have elsewhere shown. It would be rash, however, at present to speculate in this way.

Believe me faithfully yours,

RICHARD A. PROCTOR.

Proctor said "it would be rash to speculate" about the importance of the canals but he did so anyway suggesting that the lower gravity on Mars may enable "engineering works on a much greater scale than any which exist on our globe". His letter was discussed at the next day's RAS meeting. Now the cat was well and truly amongst the pigeons, intelligent life on Mars! No one had predicted it might be hostile but it was only a matter of time before they would.

Charles Darwin was not asked to comment. He woke in the night of April 15th with an unbearable pain in his chest; four days later he was dead. He is buried in Westminster Abbey alongside his inspiration Sir John Herschel.

An early example of spin

A letter written by Giovanni Schiaparelli to Otto Struve, found in 1963, suggests that the Italian who prompted one of Astronomy's greatest ever controversies, 'canals on Mars', and always denied suggesting it was anything but a natural phenomenon, may have been an early exponent of spin.

Giovanni Schiaparelli – an early publicist of martian investigation. With thanks to Richard McKim.

With his work, as it was allowed by the proximity of the Mars and Earth, finished, Schiaparelli made a presentation of his results in Rome at a meeting attended by several government officials. A few days later he was invited to repeat his lecture to the King and Queen of Italy at the Quivinal Palace. Both opportunities proved irresistible to Schiaparelli. "Mars appears to be a world little different from our own", he said, and seeing his chance added words to the effect that 'with a bigger and better telescope I could see a lot more'. When his application for funds came before the politicians, the Chamber of Deputies, the Senate and the King, everyone enthusiastically endorsed it.

"I managed the affair rather well", Schiaparelli confided to his mentor Struve in the letter to Berlin.

Apart from sparking the martian canal controversy, Schiaparelli should be remembered for a lasting legacy to Mars; it is his system of nomenclature which has been adopted.

Martian map-makers it seems are not bound by naval directives or the conventions of meteoriticists. A previous map of Mars created in the 1860s by Richard Procter attached famous people's names to various features. For example, Syrtis Major, which had also been the Atlantic canal, according to Proctor became the Kaiser sea.

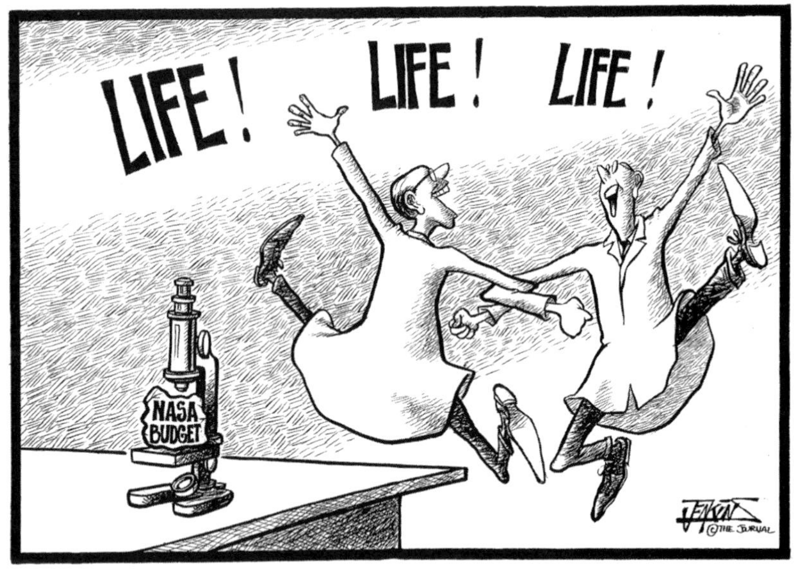

NASA was heavily criticised for its handling of the news that some of its scientists had discovered fossils in a martian meteorite in 1996. They were accused mostly of using the claim to enhance their budget. The truth may never be known but, like Schiaparelli, NASA certainly did not say 'no thank you' to the boost its Mars programme received. Cartoon with thanks to Mike Jenkins.

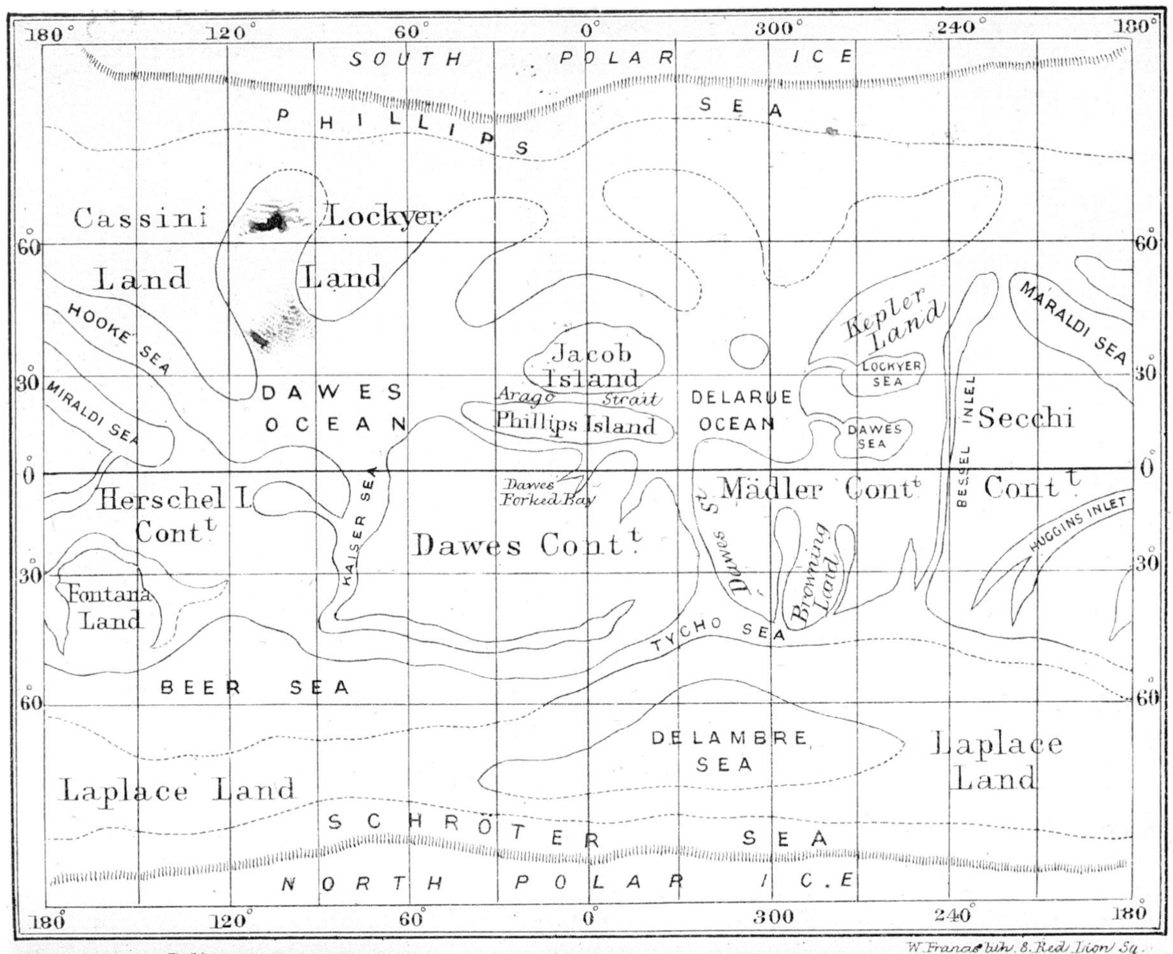

Map of Mars by Proctor, 1867, honouring a number of British astronomers. Courtesy Royal Astronomical Society, London.

Unfortunately Proctor was a little too generous with British names for his map to become popular in Europe. Schiaparelli's penchant for classical names taken from around the Mediterranean has proved enduring. Alas for the Kaiser, his sea once again became Syrtis Major. Isidis Planitia, the location where Beagle 2 will land, was given its name by Schiaparelli.

Opposing views

The best time to observe Mars with a telescope from Earth is at opposition when both Earth and Mars are on the same side of the Sun, which occurs approximately every two years two months. Because Mars has a highly elliptical orbit not every opposition is a good one. The best chances come when Mars is near Earth and close to the minimum radius of the ellipse, every fifteenth or seventeenth year.

Camille Flammarion, whose book on the earliest observations of Mars is still the authoritative work. With thanks to Richard McKim.

After the favourable opposition of 1877, Schiaparelli, with his new eighteen-inch telescope, supported by amongst others the French astronomy populariser Camille Flammarion, passionately believed he observed canals and mapped over one hundred, sometimes describing double parallel features. Wherever the channels crossed, the intersections were apparently enlarged.

But not everyone agreed with Schiaparelli. The alternative view led by Edward Maunder, President of the British Astronomical Association, claimed that canals were an optical illusion, arguing that the pro-lobby saw only what they wanted to or were told to expect. Observers could not agree amongst themselves where the canals were, let alone whether they existed. Maunder with a twenty-eight-inch telescope saw only one feature which might be a canal, whereas a pro-canals Briton, A.S. Williams, with an instrument of one twentieth of the light gathering capacity, said he saw sixty. The antis could not resist pointing out that to be observable at all the canals would have to be at least seventy miles wide! What kind of beings could dig such giant Marsworks? A popular astronomy lecturer of the time, Gresham Professor of Astronomy Edmund Ledger, helpfully suggested that because the gravity of Mars was only forty percent that of Earth, Martians would be fifteen feet tall.

Edward Maunder, President of the British Astronomical Society. With thanks to Richard McKim.

The sixth *Beagle* ship (1889–1905)

After the fifth HMS *Beagle*, the Navy must have breathed an enormous sigh of relief when the sixth *Beagle* turned out to

The sixth HMS Beagle *had a mostly peaceful career. Courtesy National Maritime Museum, London.*

be almost non-descript. She was an eight gun, twin-screwed sloop launched at Portsmouth in 1889. Her length was 208 feet whilst her beam and draught were thirty and thirteen feet respectively; she displaced 1170 tons and her 2000 horsepower engine could propel her at thirteen knots. She was sold in 1905. During her tenure of the name there were no naval battles or incidents, which was just as well there was more than enough activity concerning Mars to last anyone for a lifetime. She does however appear to have witnessed one piece of history. A diary of events written by an officer on the HMS *Beagle* recounts a South American revolution in Brazil in 1890.

Signalling Mars

Anyone opening their copy of *The Times* on Saturday morning August 6th 1892 to read the report on happenings at the British Association for the Advancement of Science would perhaps have come across a letter to the editor from Charles Darwin's cousin, Francis Galton. Galton was

Mug shots of Francis Galton who is better known for advocating the use of fingerprints in crime detection.

advocating taking advantage of the opposition of Earth and Mars to attempt to communicate with Martians by setting up a bank of mirrors to reflect flashes of sunlight to the adjacent planet.

Galton's note began by mentioning a bequest widely publicised in France. Madame Guzman, wishing to commemorate the name of her dead son, had offered one hundred thousand francs, a fortune, to anyone making contact with a planet and receiving a response. Anyway if Galton was hoping his idea was original, it was not. Charles Cross in 1869, just before the previous Earth/Mars opposition (and long before the discovery of 'canali' sparked massive speculation of sophisticated Martians) had submitted a memoir to the French Academy of Sciences suggesting how electric lights focused by parabolic mirrors could be beamed towards Mars to signal to its inhabitants.

Nobody took up the Cross proposal, maybe because there was not the added incentive of a prize. Indeed nobody decided to pursue Galton's suggestion either, but it did provoke an alternative idea from the Reverend Hugh Reginald, parish priest of St James, Marylebone. Writing to the *Pall Mall Gazette*, he suggested communicating with Mars did not need any special signalling device; it could take advantage of what was naturally available. He proposed switching all the lights of London systematically on and off at five-minute intervals to create an eye, twelve square miles in size, to wink at Mars.

He thought his scheme could be achieved with the cooperation of the gas lighting companies. The period of five minutes was chosen because it was too short for criminals to get up to any major misdeeds. Reginald's idea provoked much more correspondence than Galton's and for a few nights letters to the *Pall Mall Gazette* reminded him of everything that was wrong with his scheme, for example Martians would have to use their telescopes in broad daylight and look into the Sun to see Earth.

Whilst Schiaparelli had been reporting on Canali and everyone was arguing about Martians, Asaph Hall at the US Naval Observatory had discovered that Mars had two tiny moons Phobos and Deimos orbiting close to the planet. Nobody took much notice of Hall who could not see canals, except President Abraham Lincoln, who gave him a special citation.

Sir Norman Lockyer, editor of *Nature*, pointed out that, whilst contacting Martians was worthwhile, there were loads of cities switching on and off their lights to make Martians aware of life on Earth, without having to go to special effort.

In contrast R.H. Hutton writing in *The Spectator*, a journal much more attuned to political commentary, could not see the point. He felt that the state of development of a planet must be proportional to the amount of heat – on this score Martians, like Eskimos and Patagonians, must be less civilised than us. Assuming that Martians had eyes and enough intelligence to exchange flash for flash with us how could this possibly advance our understanding of things Martian?

Hutton was taken to task the following week by an unfortunately anonymous correspondent who explained as follows: "If Martials (sic) will and can return three flashes for our three flashes, a stupendous result will have been achieved. We have the certainty, not hypothesis, that there are other sentient beings than ourselves, that we are but units in the Universe, lost amid billions upon billions of thinkers. The horizon is inconceivably enlarged". He clearly valued highly the realisation that we are not alone; *The Spectator*'s editor thought our horizons quite vast enough already.

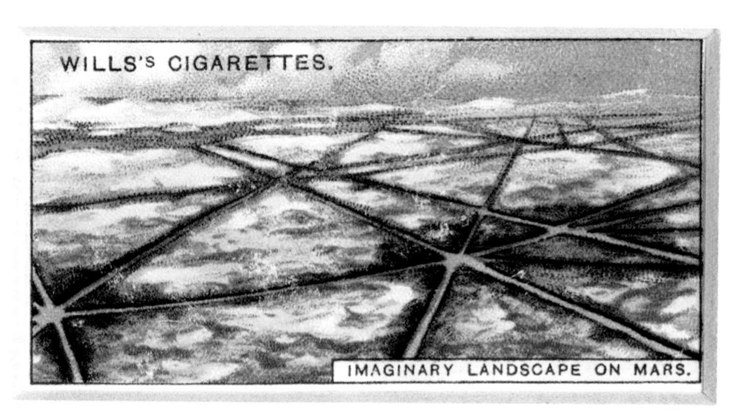

A cigarette card produced by the W.D. and H.O. Wills tobacco company of Bristol, depicting martian canals.

Percival Lowell

The explanation on the reverse perhaps suggests that Wills's were convinced.

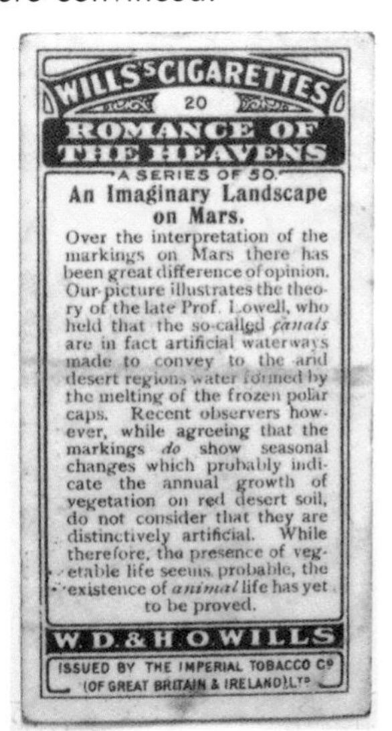

WILLS'S CIGARETTES

20

ROMANCE OF THE HEAVENS

A SERIES OF 50.

An Imaginary Landscape on Mars.

Over the interpretation of the markings on Mars there has been great difference of opinion. Our picture illustrates the theory of the late Prof. Lowell, who held that the so-called *canals* are in fact artificial waterways made to convey to the arid desert regions water formed by the melting of the frozen polar caps. Recent observers however, while agreeing that the markings *do* show seasonal changes which probably indicate the annual growth of vegetation on red desert soil, do not consider that they are distinctively artificial. While therefore, the presence of vegetable life seems probable, the existence of *animal* life has yet to be proved.

W. D. & H. O. WILLS

ISSUED BY THE IMPERIAL TOBACCO C^{O} (OF GREAT BRITAIN & IRELAND) L^{TD}

Before the favourable 1894 opposition, Schiaparelli had to admit his eyesight was failing on top of the fact he had been colour blind all along. But he was replaced by a new champion, someone who did not need the money to establish telescopes but wanted the glory – the rich amateur Percy Lowell. It is said that Lowell was switched on to astronomical observation by a copy of Flammarion's *La Planeté Mars* (probably this is still the best work for sources of early studies of our neighbour) given as a Christmas present in 1893. With the best chance to see Mars for seventeen years fast approaching, the new convert to the cause offered his services and his money to Harvard University. When it was politely declined, he poached the University's staff and built his own Observatory at Flagstaff, Arizona.

Right from the start Lowell saw everything he believed there was to see; he produced photographs which his opponents said did not stand up to enlargement. His observers catalogued 437 canals in 917 sketches from 1894 to 1916 when Lowell went to his grave still believing in Martians.

His great theory was that since Mars was smaller than the Earth, it had cooled faster therefore life developed earlier, and was now in an advanced state of evolution, fast in danger of dying out because the seas of Mars had dried up. The only water available was frozen out at the poles and each spring melt water was channelled to the fertile equatorial plains by the surviving super-beings. It was not the canals themselves that could be seen but the lush vegetation on their banks, thus the transitory nature of some of the canals and their doubling was conveniently explained; the enlarged junctions became oases.

"Artificial marking of Earth seen from Space", reproduced from Mars as the Abode of Life by Percival Lowell, 1908, The Macmillan Company, New York. Lowell argued the walkways of Hyde Park looked like the canals he observed on Mars.

He was clever – but not clever enough

H.G. Wells, in a moment of disgruntlement, wrote that his epitaph should be: "He was clever but not clever enough". Wells aspired to be a scientist, for a while he was a student of T.H. Huxley who was Darwin's champion at the Oxford BAAS meeting. But instead of becoming a biologist, he is revered as the father of science fiction. No exponent of the genre could be held in greater awe. *War of the Worlds*, sparked by Lowell's theory of a dying Mars, was one of Wells's earliest efforts. The invasion of Earth began at Horsell Common near Woking where a projectile shot from Mars landed. It was serialised in 1897 by *Pearson's Magazine* before publication in book format. Until Wells, martian wars were between those who believed in canals and those who did not.

Hyde Park and the Serpentine, London from a free balloon.

The Woking Martian by Michael Condron, 1998. The artist's impression of a Martian fighting machine, "a walking engine of glittering metal", was erected in the centre of Woking to mark the centenary of the first edition of H.G. Wells's War of the Worlds.

> "No one would have believed in the last years of the nineteenth century that this world was being watched keenly and closely."
>
> "Intellects vast and cool and unsympathetic slowly and surely drew their plans against us."
>
> *War of the Worlds* H.G. Wells 1897

Wells described the capability of invading force: "The Martians seem to have calculated their descent with amazing subtlety" and went on to explain "during the opposition of 1894 a great light was seen on the illuminated part of the disc". The narrator in *War of the Worlds* was inclined to think this blaze may have been the casting of the huge gun. The idea of Ogilivy, the astronomer in the story, was that meteorites might be falling upon the planet, he believed that "the chances of anything man-like on Mars are a million to one".

This phrase was later incorporated into the rock musical version of *War of the Worlds* by Jeff Wayne, his version was: "The chances of anything coming from Mars are a million to one, he said, the chances of anything coming from Mars are a million to one, but still they come."

The same odds could easily have been got against meteorites from Mars, but they had already come. And Ogilivy was nearly right, the meteorites are ejecta from the impact of asteroidal-sized fragments on the planet.

Name your own price

Lowell promulgated his theories in three books, *Mars* (1896), *Mars and its Canals* (1906) and *Mars as the Abode of Life* (1909). World lecture tours, and of course the papers, gave him a platform to put across his message. The opposition did not enjoy anything like the same facility to air their side of the argument.

One man, however, who was given the chance of challenging Lowell, was the self-effacing survivor of the evolution debate. Now in his eighties at the start of a new century, the co-founder of evolutionary theory, Alfred Russel Wallace, was

Alfred Russel Wallace by Thomas Sims circa 1863–1866. Courtesy National Portrait Gallery, London.

asked by the *New York Independent* to write an article for them on any subject; he could name his fee. That article, *Man's Place in the Universe* (1903), led to book reviews of Lowell's offerings and, ultimately, in 1907, Wallace's *Is Mars Habitable*? To which his answer was a resounding no. Wallace's main objection was that the low atmospheric pressure on Mars did not permit liquid water for the purpose of sustaining the canals; it was absolutely correct. His tenet that the solar system was the centre of the Universe, a view widely held by astronomers at the turn of the 20th century, was not.

The seventh *Beagle* ship (1909–1921)

The next HMS *Beagle* in the line was the biggest and most powerful yet, at 265 feet long and 12500 horsepower. She was a destroyer built at the Brown's yard on the Clyde. The last of the coal-burning destroyers, she was capable of twenty-seven knots and equipped with torpedoes to counter the first World War's new naval menace: the submarine.

The seventh HMS *Beagle*'s theatre of war was the Dardanelles, the straits linking the Aegean to the Sea of Marmara. On the north side of this three to four-miles wide stretch of water lies the Gallipoli peninsula, scene of some of the more infamous mistakes in a war of bungled campaigns.

The importance of the Dardanelles cannot be underestimated. Since myth has it that Hero was united with Leander by his nightly swim, it has been crossed and recrossed by armies going in either direction. Both the Persian King Xerxes and Alexander the Great crossed it by building a bridge of boats. At first the allies thought they

The seventh HMS Beagle, *one of eighteen Basilisk class destroyers – the whole group was sometimes referred to as the 'Mediterranean Beagles'. Courtesy National Maritime Museum, London.*

could force these straits with a fleet of rather old battleships and cruisers. The strategy failed due to the presence of mines that were laid in the channel and which destroyed one third of the fleet. The tactics then changed and an invasion force was landed to occupy the shore. It hardly got off the beaches, for a year, attempts to advance were made with fearful British and Commonwealth losses.

Amongst the ships that supported the original naval action was HMS *Beagle*. Historians have argued that if *Beagle*, and ships like her with minesweeping capabilities, had led the attack, the outcome might have been entirely different. At least the *Beagle* had survived when the futility of the effort to take the Dardanelles was realised.

The *Beagle* spent almost her entire career in the Mediterranean which is appropriate considering the next significant event in the story of meteorites from Mars.

Let the sun shine through

The problem of signalling to Mars created by the fact that the poor Martians would be looking into the Sun had been forgotten by a series of American authors by the time of the next opposition of Earth and Mars; as apparently were all the suggestions that had been made before in Europe.

In anticipation of Mars and Earth being at their closest in September 1909, W. H. Pickering resurrected a version of the Galton proposal; he could build a bank of five thousand mirrors in all a quarter of a mile square to flash a message to Mars. Some citizens of Stamford in Texas thought it was a good investment and put up $50,000 for someone to build a Mars communicator near their town. Another astronomer, R.W. Wood of John Hopkins University, however, proposed a much cheaper martian signalling device: why not simply roll out black cloth on the desert so that black and white spots could be alternated. The foolishness of this proposal, that the side of Earth visible from Mars at opposition is completely dark, was soon pointed out and the powerful light idea came to the fore again.

W.H. Pickering, the astronomer who reinvented the idea of signalling to Martians in 1909. Thanks to Richard McKim.

Given the difficulty of using light, the intrepid Professor David Todd of Amherst College announced he intended ascending to the maximum possible height in a balloon to listen out for radio signals. The grand old man of Mars science, Flammarion, could not see the need for these new fangled radiowaves. He was reported to believe it could be done by telepathic communication and what is more this would be more cheaper than mirrors.

Only a few of the budding communicators stopped to worry about how Martians would understand Earthlings' messages; at least one had the sense to point out they might not be acquainted with Morse code. Galton had thought about this in a sixty-page paper to *Fortnightly Review* in 1896. The original manuscript includes much more speculation, including how Martians reproduce, but the Victorians were a bit prim and proper so the journal declined to publish that part.

In the new round of discussions one wild idea given column inches concerned a little girl. She rationalised that a mad Martian millionaire trying to communicate with Earth would use a code depending on base eight mathematics. The little girl's reasoning was that, as an ant-like creature, he would have learned maths by counting his six legs and two antenna.

Soon the suggestions for contacting Martians degenerated into farce. A satirical article which appeared in *Science* proposed taking advantage of the Earth being between the Sun, a light source, and Mars. It required a hole several miles across being drilled to let the Sun shine through. "Although some minor difficulties remain to be overcome, many of the details are already settled," it said.

The subject was declared closed by the fictitious story of Earthlings finally making contact with the red planet through laying out a message 100 miles long using paper letters on the Saharan desert. It was said that the answer they received was "Why do you send us signals?" to which they sent the reply "We do not speak to you at all, we are signalling Saturn".

Hands off our planet

H.G. Wells killed his Martians with common terrestrial microorganisms against which they had not evolved immunity. The Martians of Percy Lowell met a more dramatic end. On the October 29th 1909, as opposition approached, under the front page headline "Life on Mars ended by a cataclysm", the *New York Times* announced that British scientists had reported that a gloomy yellow veil had spread over the planet. On an inside page, referring to the controversy of the canals, the paper said, "Well however it may have been hither to, there is no life on Mars now!"

TRYING TO STEAL OUR PLANET.

The official statement from the United States Naval Observatory by Prof. ASAPH HALL that no great changes have lately occurred on Mars places the British astronomers in an unenviable position. Having lost the north pole, the Britons have been trying to carry their national habit of seizing things beyond our planet. Mars clearly belongs to this country. That is to say, the red planet belongs to this country if it belongs to any country on earth. The statement put that way is, we fancy, incontrovertible. Only Prof. CAMILLE FLAMMARION could reasonably urge a counter-claim to ownership on behalf of his native France, and that claim would never hold in court unless Prof. FLAMMARION could actually prove his theory that Mars is inhabited by small blue persons with wings.

The English astronomers have declared that Mars has turned yellow, an announcement that savors of yellow science. They infer from this that a great cataclysm has devastated the planet. Why? Cataclysms are not often yellow. Until yesterday we had never heard of a yellow cataclysm. Mr. AUGUSTUS THOMAS in his new play, "The Harvest Moon," shows that yellow is the color of buoyancy and joy. The Martians may have been duplicating our Hudson-Fulton Celebration. From Flagstaff, Ariz., where they know all about Mars, the news comes that there is spring vapor on the planet "due to the canal development progressing according to theory."

That ought to be the final word. Great Britain must stop meddling with Mars. The sky is full of planets, systems, and constellations. Let the English pick out another planet or a whole system for their imaginary cataclysms and leave us in possession of the comparatively small sphere we have made our own.

"Trying to steal our planet" printed in the New York Times.

Lowell refused to lie down; at the invitation of the editor of the *New York Times*, he declared that "recent sensational reports had no basis in fact". This prompted the most extraordinary editorial in the paper on October 30th accusing Britain of trying to steal America's planet. It has to be read in full to be believed.

What was being observed in 1909 of course was a global martian dust storm. The astronomer Antoniadi suggested as much but the scale of dust storms on Mars was not really appreciated until Mariner 9 arrived to see, or should that be not see, the planet in 1971.

The final demise of the martian canals for professional astronomers came at a meeting of the Royal Astronomical Society on December 29th 1909. Lantern slides, of photographs taken with the most powerful telescope ever built at Mount Wilson, were shown. Edward Maunder of the Royal Greenwich Observatory said they showed no evidence of a "spider web like network". He read a report by Antoniadi which demonstrated what had looked like continuous lines were in fact unconnected features: they were an illusion. The *New York Times* called it a delusion, and quoted Antoniadi as saying "You may sleep quietly in your beds without any fear of invasion from Mars". Lowell retorted "I am hardly interested enough in the opinion of the British Astronomical Society to discuss it at length. I am however very sorry for them".

Lowell and his cohorts had done such a good job that almost every schoolboy believed in martian canals and Martians until the spacecraft Mariner 4 photographed the planet during a fly-by in 1964. Various scientists have tried to relate Lowell's canals with some real features on the planet Mars, to provide some retrospective credibility, but all their efforts have proved in vain.

As the Mariner spacecraft photographed Mars, Leslie Illingworth turned the tables on America with his cartoon of a martian canal. First published in the Daily Mail July 14th 1965. Courtesy of Associated Newspapers and the Centre for the Study of Cartoons and Caricature, University of Kent.

A dog's life

At the Eastern end of the Mediterranean, the Nakhla stone which fell in Egypt in 1911 is famous, or infamous, on several counts. It is our largest martian meteorite at about fifty kilograms and is said to hold the distinction of having been involved in a fatality: it is supposed to have killed a dog. The supporters of Beagle 2, like the readers of the *Egyptian Gazette*, which brought Nakhla to the attention of the world, will no doubt be distressed to learn that a fragment "fell on a dog at Denshal leaving it like ashes in a moment". This sounds highly far-fetched. That the dog could have been hurt, even killed, by a several kilogram object at terminal velocity is not in question but the suggestion that it could have been hot enough to cause spontaneous combustion of the animal is a bit over the top!

Polished thin section showing the crystalline structure of the Nakhla meteorite.

Studying carbon in meteorites

Whilst it might not have caused a fire, a sample of Nakhla has been analysed by combustion in the laboratory. Stepped combustion, to be more precise, a process whereby the sample is heated in oxygen over a series of temperatures to see what burns; different forms of carbon are identifiable by their relative stability. Applied to Nakhla, the technique suggests the presence of a component of carbonate, an interpretation which was verified by showing that it did not need oxygen to be present for decomposition to occur. Similarly this carbon-containing material was removable with dilute acid again suggesting its existence as carbonate. The carbon isotopic composition of the carbonate showed it must have been formed from interaction between the martian atmosphere and the rock.

Now this is an unexpected result since Nakhla is an igneous rock (formed volcanically) and not the sort of specimen that should contain carbonate. When it was announced that geochemists had found carbonate the discovery was greeted with disbelief by petrologists who had spent years looking at polished thin sections of rock and not seen any evidence of the mineral.

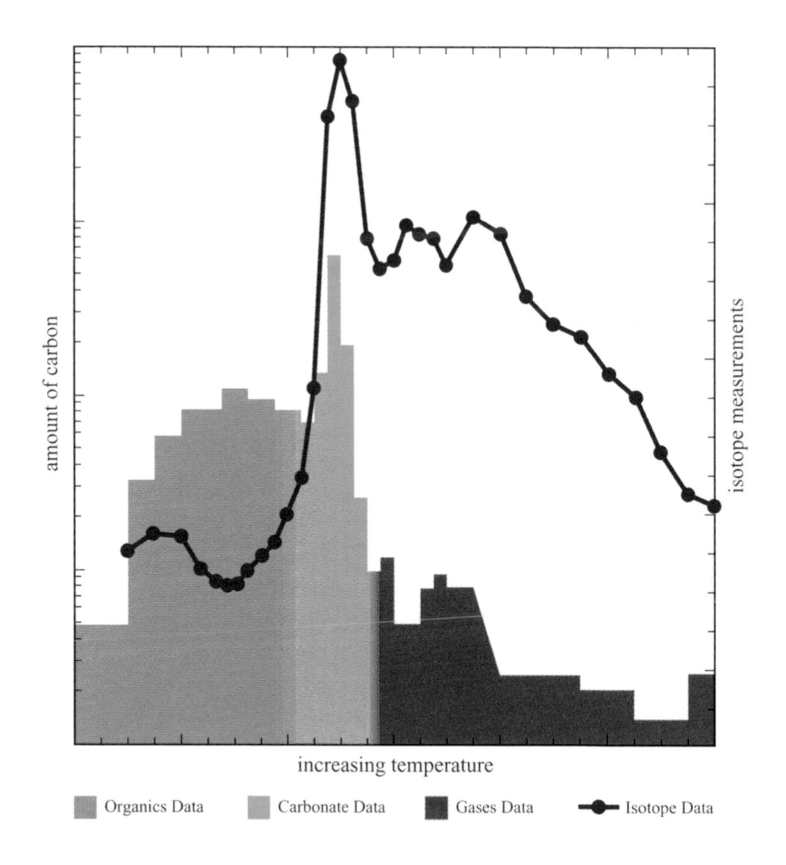

Stepped combustion diagram for ALH 84001, a very important meteorite. The histogram shows the amounts of carbon at each temperature step, different colours correspond to different kinds of carbon, distinguished by the temperature at which they burn. The individual points joined by the line are a measure of the isotopic composition.

Radio Ga-Ga

Even though the astronomers had settled their differences, the communicators had not. The idea of communicating by the wireless telegraph was still on the agenda. At first it had been thought impossible. Marconi's fabulous new invention initially had a range of fifty miles Then it was found that messages could be sent from one ship to another even if it was below the horizon. Then momentously a radio signal traversed the Atlantic. But still it was thought that the thirty-five million miles between Earth and Mars required a transmitter on a scale hopelessly beyond the power of man. "The subject of alleged communication with Mars therefore seems to me to be undeserving of serious attention", one critic wrote.

The song Radio Ga-Ga by the band Queen refers to the radio version of H.G. Wells's War of the Worlds.

Unfortunately, that was not the end of the matter. J.H.C. MacBeth, a London manager for the Marconi Wireless

The Needles headland on the Isle of Wight from where, according to a plaque on the site, Marconi made the first transatlantic broadcast. Other locations also claim the honour. It may be that the first shore to ship broadcast was made from this site. Here the British rocket programme started and ended. The concrete remains are all that is left of the gantries in which, from 1956 to 1971, Black Knight and Black Arrow rockets were held down whilst their engines were test fired.

Company, speaking at a Rotary Club luncheon, held at the McAlpin in 1920, claimed that the company had succeeded in making contact. Mr MacBeth said "he could not believe it would prove too difficult to interpret messages, after all, during the recent war, no code had defied breaking for more than about three weeks". He obviously did not succeed as that was the last heard from him.

The eighth *Beagle* ship (1930–1945)

The eighth HMS *Beagle* was another destroyer, a B class vessel, this time of 1360 tons. She was again built in the shipyard of John Brown on the Clyde. As destroyers went, the *Beagle* was small; other nations with whom Britain would find itself in conflict were building much larger vessels but Britain felt bound by naval limitation treaties.

HMS Beagle, *H30, in Torbay on July 10th 1937. Courtesy National Maritime Museum, London.*

Overall HMS *Beagle* was 323 feet long and almost exactly a tenth of that dimension at the beam. In terms of armaments she carried four 4.7 inch guns and five 3.3 inch anti-aircraft machine guns. There were two banks of four torpedo tubes, each torpedo being twenty-one inches in diameter.

The eighth *Beagle* began her life in a time of peace but the threat of war loomed large in Europe and thoughts of invasion even crossed the minds of Americans.

Invaders from space

If Pasteur had written up his work on looking for microorganisms in meteorites then Charles Lipman in 1936 would not have been able to claim "I was and am the pioneer on this subject...". Most scientists are more modest but Lipman was under the impression he had successfully identified living bacteria from space.

If he did not know already about the work from the 1870s, then it was a fairly safe bet that such an arrogant statement would prompt someone to tell him that Pasteur had some claim to precedence in respect of being the first to consider the possibility of meteoritic life. So Sharat Roy not only informed Lipman he was wrong but also told him others had been wrong before him.

On the second count, only the anonymity of having written in French spared the blushes of Monsieur Gillipie and Madame Souffland in respect of their positive results acquired between 1911 and 1921.

Whilst Lipman of the University of California fought it out with Roy, his arch-enemy from Chicago, with the *New York Times* looking on, the French pair remained remarkably silent considering they had said, in 1921, that their

THE STAR, THURSDAY, OCTOBER 28, 1926.

IS THERE ANY LIFE ON MARS? CERTAINLY.

Even though professional astronomers no longer believed in Martians, Percy Lowell's ideas had spread far and wide and the public had other perceptions as this cartoon by David Low shows. First published in The Star *on October 28th 1926. Courtesy of Associated Newspapers and the Centre for the Study of Cartoons and Caricature, University of Kent.*

experiments (which included the martian meteorite Chassigny) "demonstrated meteorites whatever their composition contained organisms susceptible to being revived and multiplied".

The second War of the Worlds

In 1938 radio waves finally contributed to the martian debate. In the early days of broadcasting, families used to gather in the evening to be entertained by the box in the corner. There was a good deal of channel hopping to find a new offering particularly if the programme on one station did not come up to expectations. Retuning those old fashioned radios took time so frequently the first few minutes of a programme were missed.

On October 31st, a large fraction of the audience who joined CBS's New Mercury Theatre, some because they had changed from a radio ventriloquist, did not hear the warning about a play that was to follow. When some light music they were hearing was interrupted for a special news announcement, they thought it was just that. But it was not. It was an attempt by Orson Welles, or so he said, to spice up the H.G. Wells classic tale of the Martians landing in the Thames Valley. With Nazi storm clouds gathering in Europe, and invasion an everyday word, the USA was ripe for being scared out of its living day-lights.

"Hurry! They'll be turning on their cameras any second ..."

Cartoon by Bernard Cookson, first published in the Evening News, *July 30th 1969. Courtesy Associated Newspapers and the Centre for the Study of Cartoon and Caricature, University of Kent.*

Orson Welles had time-travelled the sci-fi classic to the 1930s, to Grover's Mill, New Jersey, and added outside broadcast realism. Soon his listeners were being treated to breathless eyewitness reporters describing scenes where Martians, impervious to gun shots, played with their victims like a cat and mouse. With the CBS switchboard jammed, mid-broadcast exhortations that it was only a play were taken as a Government attempt to cover up and allay panic. Anyway many had already taken to the streets, searching for or trying to reach loved ones.

At the end of the broadcast the Producer/Director reminded the listeners it was, after all, Halloween. Next day papers were still trying to restore calm. Orson Welles always denied it was premeditated to boost ratings but it certainly never damaged his career as Hollywood soon beckoned.

Battle honours

More than half the battle honours accumulated by the pack of HMS *Beagles* can be attributed to the eighth vessel of the name. She fought off the coast of Norway and the Bordeaux region of France in 1940, evacuating British nationals at the start of the second World War. She then patrolled the Dover Straits as the Navy's first line of defence against invasion, being damaged by dive-bombers but giving as good as she got in shooting down an attacker.

Most of the eighth *Beagle*'s war however was spent on convoy patrol, the majority of the time in the western approaches of the North Atlantic shepherding cargo ships to and from Iceland. She went eight times to Russia, to Murmansk, on the desperately dangerous Arctic runs, making several trips before she was 'arcticised'. With typical British idiosyncrasy, once equipped to fight the cold she was despatched to Freetown, West Africa because of the increased U-boat activity in the tropical area.

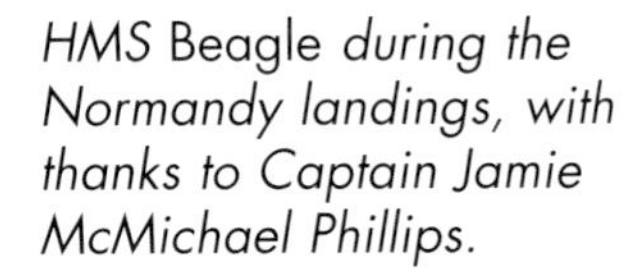

HMS Beagle *during the Normandy landings, with thanks to Captain Jamie McMichael Phillips.*

Her Russian convoy exploits included fighting off an attack in which she was considerably outgunned by an enemy force; the Germans possessed fifteen five- to six- inch guns, whilst the British contingent could muster only nine smaller calibre weapons between them.

On another occasion this *Beagle* played a part in sinking a U-boat with depth charges although the success was not appreciated until after the war was over and the loss of the U-355 in the action was discovered.

With the tide of the war turning, HMS *Beagle* took part in *Operation Torch*, the invasion of north Africa. On D-day, June 6th 1944, she escorted twenty-two landing craft to Juno beach. Two days later she rescued 250 American Marines whose landing craft had been sunk during the run into the shore. Thereafter she could be found bombarding installations on the Atlantic coast of France, occupying German forces who might have reinforced against the Allied invasion.

On May 9th 1945, a British force liberated Jersey and the German Commanding Officer was taken aboard HMS Beagle *where he signed terms of surrender. The edge of the commemorative coin bears the inscription recalling Winston Churchill's words, 'our dear Channel Islands will also be freed today.'*

The highlight of the eighth *Beagle*'s war came on May 9th 1945 when she was involved in freeing the Channel Islands. Not a shot was fired, the most defiant act being by a German officer who refused to surrender but was talked round. On the 12th she ferried Brigadier Snow from Guernsey to Jersey to read the proclamation of liberation. Just twelve days later the noble vessel was declared surplus to requirements and transferred to the breakers yard at Rosyth.

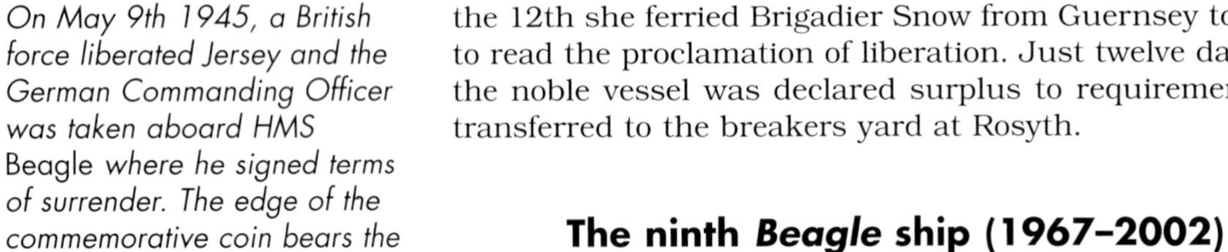

The ninth *Beagle* ship (1967–2002)

Twenty years after the end of the Second World War a new HMS *Beagle* emerged in a very familiar guise. Built by Brook Marine at Lowestoft and designed to operate in coastal

HMS Beagle*, with thanks to Andy White, Hydrographic Survey Squadron photographer. Crown Copyright.*

waters, she was a Bulldog class survey ship, thereby emulating her famous ancestor, the third *Beagle*. She was, however, twice the size of FitzRoy's command, measuring 190 feet, thirty-eight feet at the beam and displacing 1070 tons. The crew was about half the number of the 1830s ship – six officers and thirty-six men. How Darwin would have envied their passive stabilisers, full air conditioning and individual accommodation. There was even a well-appointed sick bay he could have put to good use. For her special hydrographic role the ninth *Beagle* was equipped with a variety of sonar and radar equipment. Even in this day and age there are hazards to shipping to be mapped, not least in the Persian Gulf, the Caribbean and around the coast of Great Britain.

The ninth *Beagle* admirably maintained her link with the past through her glass-fibre survey motorboats called appropriately, *FitzRoy* and *Darwin*; the Landrover for on-shore activities had no equivalent in Darwin's times other than locally-hired real horse power for treks across the South American continent. Like the third *Beagle* the ninth one was

The author during a visit to the ship, with thanks to her Captain and crew.

equipped for spending long times at sea in isolated areas having a range of 4500 miles at twelve knots.

Desolate, depressing, sterile and repulsive

At least the Mariner and Viking results killed off one Russian theory. In 1959 the Soviet Youth Newspaper, *Komsomolskaya Pravda*, citing Dr L Shklovsky, claimed that the martian moons Phobos and Deimos were artificial satellites, launched by spacefaring Martians. Dr Shklovsky even believed he knew when – sometime between Herschel failing to find martian moons (1787) and Hall succeeding (1877).

These words were used to describe Tiera del Fuego, Darwin's immediate destination in the third HMS *Beagle*. Charles Darwin wrote to his sister Caroline on April 6th 1834 that if anyone caught him going back to Tiera del Fuego "I will give him leave to hang me up as a scarecrow for all future naturalists". Hardly the place then to learn about the origin of life on Earth, but as everyone knows in this remote corner of the world were sown the seeds which led to one of the greatest works in biology.

The same words could undoubtedly have been used about Mars when scientists studied the first pictures of the martian surface supplied by the Viking landers. They had looked at the twenty-two images of a crater-pocked surface provided by Mariner 4 in 1964, and the vast chasms, huge volcanos and the planet-wide dust storm seen by Mariner 9 in 1969, but nothing had prepared them for the red desert before their eyes in 1976.

Over a billion dollars had been invested to look for evidence of life using a biological package of equipment on board the Viking spacecraft. The experiments however were almost entirely designed to reveal actively metabolising organisms on the red planet. There was one test which could measure gases above a soil sample when moistened, another which looked for radioactive emissions when soil samples were

The desolate surface of Mars seen from the Viking lander. Courtesy NASA.

added to a nutrient broth of radiolabelled compounds, and a third which exposed the soil to a radioactive gas mixture simulating the martian atmosphere to investigate whether any assimilation occurred.

Within days of landing on Mars all of these experiments had given apparently positive results. But still the scientists held back from declaring the discovery of martian life. They had a major problem, a fourth experiment designed to detect organic matter in the martian samples, analysis using a gas chromatograph–mass spectrometer, failed to find any organic compounds, other than some tiny residuals of solvents used to clean the apparatus.

The experimenters were perplexed: it looked like they had indications of life but could not find a body, either alive or dead. Since only science fiction writers believe in silicon-based life, the conclusion had to be that the martian surface was playing an enormous confidence trick on the patiently waiting scientists and media. The red colour of the soil was a function of the oxidising conditions on Mars. The chemical reactivity of the martian soils was mimicking the life process.

The scientists did not actually say there could not be life on Mars, only that they had not found it. In science, the absence of evidence is not evidence of absence.

Everyone fully expected that NASA would be back in the near future to collect martian materials and bring them home for study in Earth-based labs. But that did not happen, the politicians saw it differently – no life, no money, no more missions.

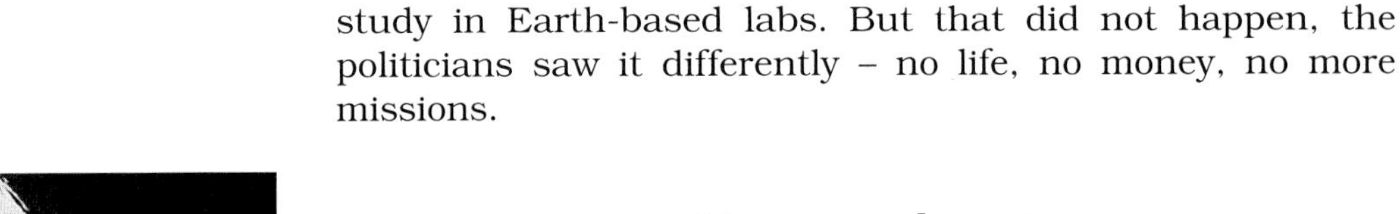

Mars and the Moon at a time of occultation (one heavenly body concealing another). Studies of rocks from our satellite were key to realising that martian meteorites existed on Earth (photograph from the Lowell Observatory at Flagstaff).

Young at heart

As a result of the availability of Apollo Moon rocks, from 1969 onwards it became possible to work out accurately the ages of extraterrestrial samples in the laboratory. Rocks similar to the strange Indian sample of Shergotty had been found on Antarctica. These, together with the French rock Chassigny and the Egyptian meteorite Nakhla were dated. All had one thing in common, incredibly low ages. Logical scientific thinking began to piece together a scenario. Moon rocks, crystallised during the heat of volcanic activity, had been found but none younger than about three billion years had been discovered, meaning that the Moon must have stopped being an active body long, long ago. All known asteroids are smaller than the Moon so they ought to have cooled down earlier. A source of more recent volcanism, that is giving rise to much younger rocks, had to be the likely source of the **S**hergotty-like material, **N**akhla and **C**hassigny, now affectionately lumped together as the SNC meteorites. A bigger body which could have had volcanic activity more recently than the moon could only be a planet – could it be Mars?

The football-sized meteorite EETA 79001 showing the black glassy inclusions. NASA image.

A martian fingerprint

A Shergotty-like meteorite (they are all called Shergottites) found in a region of Antarctica called Elephant Moraine (shortened to EETA) in 1979 provided the answer to the question, 'Could meteorites come from Mars?' A scientist working in Houston melted glassy parts of EETA 79001, an eight kilogram football-sized object, and showed that the isotopic compositions of noble gases (argon, krypton and xenon) released suggested a mixture of components, one having the Earth's atmospheric composition and the other similar to the atmosphere of Mars as measured by the Viking lander on the red planet. The only way this could be was if the glass in meteorite had trapped tiny bubbles of the martian atmosphere when it melted during the impact which threw the meteorite off Mars. It was a great idea but it relied on a knowledge of the trace constituents of the martian atmosphere only. Then UK scientists using a mass spectrometer, newly developed to study minute samples of carbon dioxide, extended the study and showed that EETA 79001 contained carbon dioxide, the major constituent, ninety-six percent of the martian atmosphere, in the correct proportions.

The match between the atmospheric composition of Mars and the gas in the meteorite became known as the 'fingerprint' of Mars. EETA 79001 (and the other Shergottites) were shown to be martian rocks.

The implications were obvious

By 1989, the idea that martian meteorites existed on Earth was well established. Similarly, the petrologists had microscopically searched through polished thin sections and agreed that their igneous rocks did contain carbonates. In fact EETA 79001 had some small but clearly recognisable carbonate inclusions. When dug out of the sample and burned using the stepped combustion method, these inclusions showed the presence of significant quantities of organic matter.

On Earth, rocks that contain carbonates are amongst the richest in organic matter. Over fifty percent of them are tapped as sources of petroleum. The juxtaposition is not an accident, carbonates are precipitated from aqueous systems and the organic components that constitute oil are deposited as the debris of once-living organisms that existed in the water or were carried there by its action.

A little piece of heaven. Dr Ian Franchi contemplates a meteorite found on the Antarctic ice cap. Over 20000 have now been recovered, a few of which are from Mars.

The fact that the martian meteorite contained a small indigenous component of organic matter was published on the 20th anniversary of the day Neil Armstrong first walked on the Moon. The paper said the organic matter could have been added to the meteorite from extra-martian sources but if it was formed on Mars then "the implications were obvious".

The signs of life on Earth

Not all terrestrial sedimentary carbonate-containing rocks are petroleum source rocks but they almost all contain some organic material in addition to the inorganic carbonate of the rock. Whenever the two carbon-containing entities are found together, they have different carbon isotopic compositions, an observation attributed to the fact that biology has a preference for the lighter mass (carbon-12) isotope of carbon. Isotopes are different forms of an element having the same atomic number (number of protons) but different atomic weights (number of neutrons). Studies have been made of rocks from every geological epoch on Earth, including the oldest known sediments on the planet and the result is always the same. The organic matter has about one to three percent more of the carbon-12 than in the associated carbonate. This is taken as evidence that life on Earth began nearly four billion years ago, just as soon as the planet was stable enough to support it.

84001

What was not known in July 1989 was that another key martian meteorite had already been collected.

ALH 84001, a grapefruit-sized rock, was found by a NASA scientist during the 1984 field season at the Allan Hills in Princess Victoria Land, Antarctica.

ALH 84001, perhaps the most famous martian meteorite of all. NASA image.

Experienced meteoricists can often tell immediately when they have encountered one of the really interesting meteorite samples to be discovered on the polar ice cap. Such specimens are noted and they get opened, numbered, and more carefully assessed when an Antarctic shipment arrives home. This is the reason why 79001 and 84001 both ended up as the earliest samples appraised for their season, not because they were the first rocks found that year.

But back at the meteorite processing facility in Houston, Texas, ALH 84001 was, after all, not thought to be particularly interesting and it was packed away, unstudied for a further eight years.

When it was issued for an unrelated project, its martian provenance was recognised.

More exciting still, ALH 84001 was a very unusual martian sample, it tended to break open along various cracks exposing veins of a strange orange-coloured carbonate.

A test for martian rocks

Not all martian meteorites contain the trapped atmospheric gas that identifies them as martian. Fortunately there is another way of recognising samples. It involves measuring, very carefully, the proportions of the three isotopes of the element oxygen – oxygen-16, 17 and 18. For rocks on Earth (or for meteorites that come from another part of the solar system) the results for samples that are related show a distinct pattern, a slope of 1/2, when isotope data are plotted on a graph comparing a measure of 17 to 16 against 18 to 16. Each different provenance has a unique line; one exists which has EETA and other Shergottites, Nakhla and Chassigny on it. Using this test it is now known that there are about thirty martian samples, one of them is the most famous martian rock of all, ALH 84001.

The orange carbonate rosettes on a broken surface of meteorite ALH 84001, courtesy Dr Monica Grady.

Fizzy water on Mars

The orange veins in the ALH 84001 meteorite removed the last vestige of doubt about martian carbonates. Scientists could see it, touch it and measure it to their heart's content. The bulk carbon isotopic composition of the stuff was way different from anything similar on Earth; the orange carbonate had to be from where the meteorites came from – Mars.

With such large amounts it was possible to make other analyses, for example of the oxygen isotopes existing in the sample. The result implied that the carbonate was precipitated from a solution made by dissolving the carbon dioxide of the atmosphere of Mars in water, just like carbonation to give sparkling water. So, there was warm liquid water or as one newspaper put it "Fizzy Water" on Mars. The universal solvent, the magic ingredient for life, existed.

A martian can of worms

Worms intrigued Charles Darwin; only a year after his return from South America, much to the consternation of his audience, he lectured on them to the Geological Society; the audience probably had been expecting a more substantial subject.

Writing about worms was Darwin's last act of authorship and when *The formation of vegetable mould through the action of worms, with observations of their habits* appeared in October

1881, it was a phenomenal success – thousands of copies sold in weeks!

"Worms have played a more important part in the history of the world than most persons would at first suppose."
Charles Darwin concluding his manuscript on worms, October 1881.

What fascinated Darwin was the worm's apparent intelligence; he noticed that in dragging leaves into their burrows they always did so by the narrowest end first. It was clearly not a trial and error process. To aid in his research Darwin set to work filling the Down House greenhouses, his study, and even his precious billiard room with pots and jars to stimulate typical worm populations which he believed to be 53,767 to the acre. He tested his worms' reaction to sound, shouting or blowing whistles and getting his family to play their musical instruments, such as the piano and the bassoon, to obtain a response, but the worms appeared impervious.

Light, on the other hand, produced an immediate response, only the worms' enjoyment of sex overcame their fear of a strong light suddenly shone on them. To study their reaction to various wavelengths of light, Darwin's family tolerated him blundering around the pitch-black house bent on conducting some new observation using candles, paraffin lamps with red and blue filters and even the heat cast by a glowing poker.

No doubt Darwin would have been absolutely delighted with the next development in the martian meteorite story.

A fossil worm

Despite the harsh treatment meted out to Otto Hahn, fossils in meteorites have been a subject of debate on numerous

The author in the garden at Down House in Kent where Darwin lived after his voyage on HMS Beagle and where he did his studies on the humble earthworm.

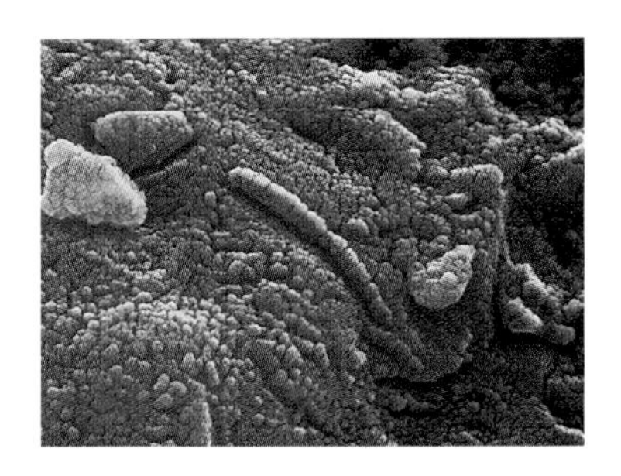

The putative fossil from ALH 84001; its size just 100 nanometres long. Courtesy of Dr Everett Gibson.

At 2.15am on August 7th 1996 the author's telephone rang. At the other end was a journalist who breathlessly enquired "is it true this rock is full of worms?" The journalist did not use any of the quotes he was supplied with, perhaps they were all unprintable!

occasions since. In the 1960s fossils or organised elements, a less evocative term, were discussed in respect of carbonaceous chondrites. The same subject was rediscovered in the early 1980s.

It was again a talking point in the coffee areas of the 1996 meeting of the International Meteoritical Society in Berlin. Rumours were rife, something had been discovered in a martian meteorite found in the Allan Hills mountains of Antarctica.

A hint of the findings appeared in the trade newspaper, *Space News*, in August. Then a BBC Science Correspondent filed a story without really any concrete information for the evening news on August 6th. He claimed fossils had been seen in martian meteorite ALH 84001.

Stunning

At 6.00pm British time on August 7th Bill Clinton became the second US President to pronounce on the subject of stones from the sky.

Making a statement as he boarded a helicopter on the White House lawn, he hailed the discovery of meteorite fossils as stunning: "The rock 84001" he said "speaks to us across all the billions of years and millions of miles. It speaks of possibilities of life. If this discovery is confirmed, it will surely be one of the most stunning insights into our Universe that science has ever uncovered".

He could have added on a par with the writing of *On the Origin of Species.*

But as he paused, obviously willing to field a question or two about how the USA had discovered life on Mars. Instead he was asked a completely unrelated political question. Now that really was stunning!

Life in the solar system and beyond

Controversial as the martian worms have become, scientists had every right to look for fossils in ALH 84001. Most meteorites come from asteroids, small dry, airless bodies, so fossils, indicative of evidence of life, would not be expected. But Mars is a planet, a planet with an atmosphere of carbon dioxide. It demonstrably had a hydrosphere in the past and water still exists according to the meteorite studies, so why not a biosphere? There are many people who would be happy to believe that Mars, smaller than Earth and farther out from the Sun, cooled quicker and was stable enough to support life earlier; and not just Percival Lowell either. ALH 84001 is a very old rock, it crystallised billions of years ago. The hypothesis which arose from the discovery of the fossils was that perhaps life on Mars began to evolve far back in the past, but when the water on Mars dried up three billion years ago, it died out.

A cartoon by MATT published on the front page of the Daily Telegraph on the morning of November 1st 1996 alongside a story concerning work on EETA 79001. Courtesy of MATT and the Telegraph Group Limited (1996).

It was contentious but not quite the same as Martians. Fossils don't mean life right now to most people. On October 31st, Halloween again, the world was reminded of the parallel work on martian meteorites done on EETA 79001.

Unlike ALH 84001, the shergottite EETA 79001 is a very young rock, in fact the youngest known martian sample, no more than perhaps 180 million years old. The alteration to produce carbonates had to have taken place sometime after formation, but before the rock left Mars. Since it can be shown that ejection from Mars was six hundred thousand years ago, the carbonate could have been deposited at anytime up to the last minute, a time when our ancestors were walking about on Earth.

So if the organic matter in EETA 79001 is the relict of organisms that lived in the water which percolated through the meteorite to produce the carbonates then it could mean life is still going on on Mars right now.

Exobiology, the study of life outside our World, took on a new lease of life. The next day a committee formed by the European Space Agency began to discuss to possibility of life on Mars.

Not quite the last of the line

On her last deployment in 2001, the ninth HMS *Beagle* sailed the equivalent of around the world thereby upholding the tradition of her namesake of the 1830s. In an active life of thirty four years *Beagle* travelled a distance of twenty one times around the planet.

On the final voyage home there was still room for a last act of bravery. The crew responded to a Mayday call on New Year's day 2002. The MV *Aydin Kapitan* found itself without power and drifting in mountainous seas. The arrival of a Royal Navy boarding party of three, in one of *Beagle*'s tiny boats, was greeted by the 230 people on-board as their salvation.

Decommissioning of HMS Beagle. *Photograph from* Navy News.

Lt O'Sullivan, P.O. Hawksby and M.E.M. Burgess were decorated during *Beagle*'s decommissioning ceremony for their role in getting the *Kapitan* under control and its engines restarted.

Although HMS *Beagle* has now been decommissioned, she has not been scrapped. She is being refurbished as a luxury yacht. She cannot of course carry her illustrious name – that for the time being will live on with Beagle 2 on its mission to Mars.

On an earlier visit to HMS Beagle *with the lander model in the chart room.*

Tales of two Beagles

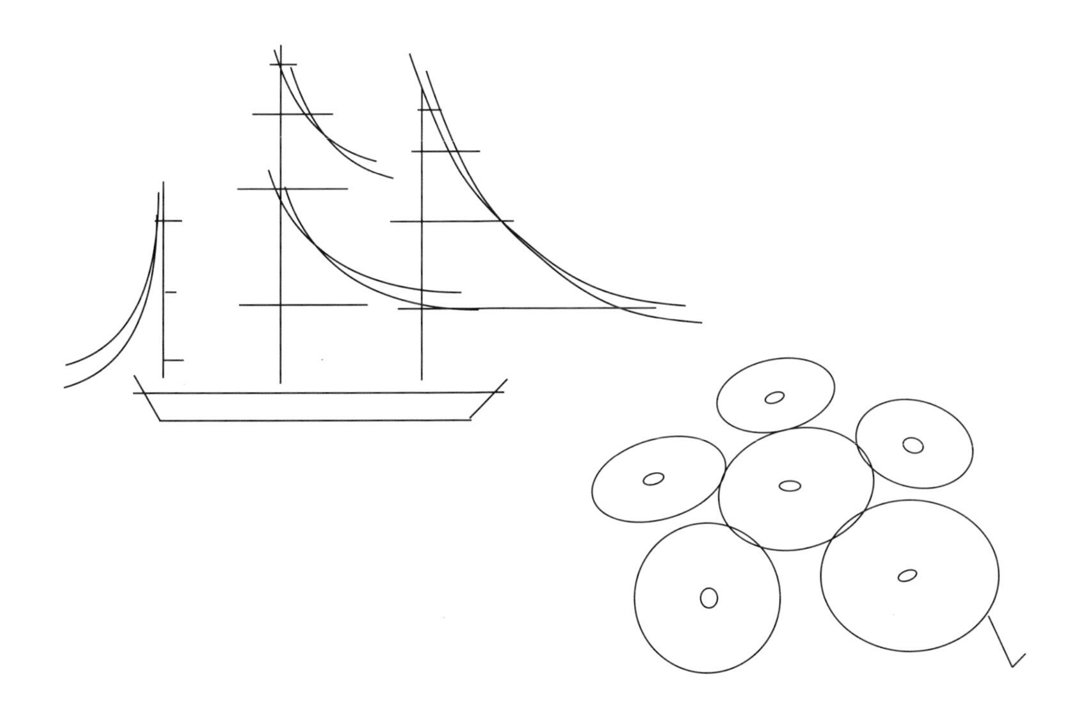

A detailed examination of the many parallels between Beagle 2 and its most illustrious predecessor, the third HMS *Beagle*, comparing one great age of discovery with what could be the next.

A cluster of failures

The opportunity for Beagle 2 arose in a place that the third HMS *Beagle*, Darwin's ship, would have known well.

With all the optimism of peristroika and glasnost, European countries, particularly France, Germany and Italy had arranged to provide instruments for a Russian space mission to Mars. The launch was planned for November 17th 1996, everything went according to plan until it became time to ignite the fourth stage of the rocket to push the Mars 96 mission into a highly elliptical solar orbit which would transfer it to Mars. Unfortunately the opposite effect was achieved and the whole lot returned to Earth, the errant booster crashing into the Pacific Ocean just off the coast of Chile.

It was a bad time for European space science. Another mission, one of the European Space Agency's cornerstones, four satellites called Cluster, intended to examine 'space weather', had also been lost when, for the first flight of Ariane V, Ariane IV software was loaded. As a result the rocket set off like a demented firework and had to be aborted.

The coast of Chile by Robert FitzRoy. Courtesy UK Hydrographic Office, Taunton. Crown Copyright.

"Bad news, Comrade President - our Mars probe has failed"

Cartoon by Mac [Stan McMurtry] first published in the Daily Mail, November 18th 1996. As the Russian Mars 96 probe plummeted to Earth, the country's leader Boris Yeltsin was in hospital recovering from a heart operation. Courtesy Associated Newspapers and Centre for the Study of Cartoons and Caricature, University of Kent.

Phoenix from the flames

Just as the illustrious career of Darwin's *Beagle* was almost stillborn, by being mothballed shortly after launch, so too were Europe's ambitions for Mars exploration. But with its dreams in tatters, ESA had been considering how to resurrect its expensive Cluster mission and had found a way it could afford by using a relatively cheap Russian rocket called Soyuz-Fregat for the launch. In 1997, it was realised that the same launcher could be employed to send a spacecraft to Mars. What is more 2003 was going to be the best opportunity for perhaps twenty thousand (later revised to nearer seventy thousand) years. The proximity of Mars and Earth during the 2003 opposition allows the maximum payload to be transported to Mars for the minimum expenditure of fuel.

There was no time to lose, the spacecraft to take the spare instruments from Mars 96 would need to be built in record time, hence ESA's Mars recovery mission would become known as Mars Express, or MEx for short.

HMS *Beagle* – grandson of Mars

Despite their unfortunate sobriquet of 'Coffin Brigs' the Cherokee class of vessels must have been a huge success, over a hundred were built. One of these was a ship called *Cadmus*. In Greek mythology Cadmus was married to Harmonia which made him a son-in-law of Ares, or as the

A Soyuz launch successfully carrying two of the Cluster II satellites into space, July 16th, 2000.

Romans called the God of War, Mars. Cadmus was responsible for bringing the alphabet to Greece and sowing the seeds which began the warrior tribe of Sparta. The instructions from the Admiralty to build the third HMS *Beagle* are written on the back of the original plans for *Cadmus*.

HMS Cadmus *circa 1821, courtesy National Maritime Museum, London.*

A model ship

Nobody made a model of Darwin's ship HMS *Beagle* before the real thing was built. A number of attempts have been made in modern times to gain an impression of what she was really like. Model makers and artists alike were thwarted by the lack of drawings for HMS *Beagle* in her final form. Considering that she had on board a variety of artists, very few contemporary paintings of the ship exist. Thus it was to the extensive writings by many on the ship, including Darwin himself, who recorded all he saw, that others later turned to tease out the details of her design.

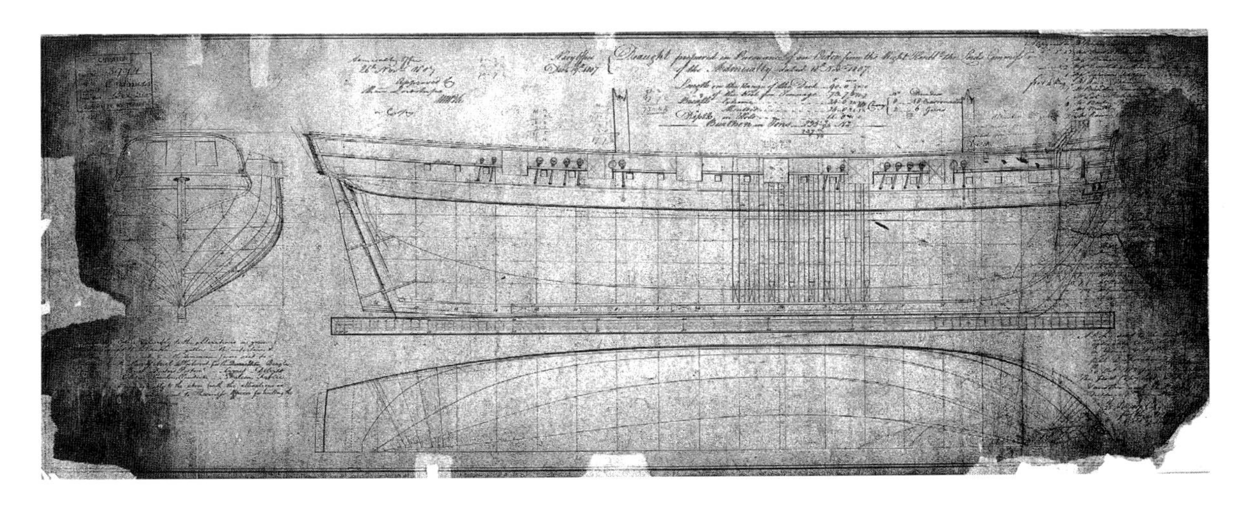

HMS Cadmus *plans, courtesy National Maritime Museum, London.*

The author with a scale model of HMS Beagle *pictured at Down House, with thanks to Down House and English Heritage.*

Instructions to build a model Beagle can be found in the *Anatomy of a Ship – HMS Beagle – survey ship extraordinary*, by Karl Heinz Marquardt.

Artist John Chancellor was another who gathered together all

HMS Beagle *in the Galapagos by John Chancellor. This fine study is a reconstruction showing the ship sailing to the far point of James Island. Using the original charts and available records, the artist visualised* Beagle*'s position on October 17th 1835. With thanks to Mrs Chancellor for permission to reproduce.*

the information he possibly could in order to paint HMS *Beagle* as she might have appeared in the Galapagos Islands.

A model space project

The first model of Beagle 2 was made of cardboard. It proved invaluable in demonstrating just how small the Mars lander would be.

The cardboard model saw action at many demonstrations including TV programmes like *Blue Peter* and the *Big Breakfast* with Johnny Vaughan.

Johnny Vaughan, Channel 4 TV: "...25M [pounds] and we can send this little babe to Mars"
Author: "Yep"
Johnny: "it's only made of cardboard"
Author: "British ingenuity"
Johnny: "British ingenuity – the first to send a cardboard probe"
Author: "Well when we get the money we will make it out of something more substantial"

More models swiftly followed. A space mission is all about making models to test all aspects such as mass, centre of gravity and electrical interfaces, demonstrating that all components work at all stages during design and development.

In an ideal world, Beagle 2 would have been built in a logical sequence with test programmes completed before moving on to the final assembly of the lander and probe that would be sent to Mars. Unfortunately, there was not time to do everything in series, before early 2003, when Beagle 2 joined Mars Express for its journey to Baikonur, so a parallel programme of working had to be arranged. Versions (models) of Beagle 2 had to be made available with the various functions needed. Different sections of the team were then able to proceed flat out with their own aspect of the work.

Concept models of Beagle 2, 'Blue Peter Style' made with cardboard, coloured paper and paper towel rolls. The second version had corners removed to facilitate greater movement for the arm.

Some would not be recognisable as anything like the full-size model of the lander used for public exhibitions. For electrical tests, it would be a box of electronic circuit boards wired together; sometimes the box was left out. Other models looked more representative, but all had their purpose.

The Development Model (DM) is the name given to any version used to test a function of Beagle 2. For example, DM gas-bags were used for tests to demonstrate the design and capabilities of the system which protects Beagle from the impact of hitting Mars. An Aeroshell Assembly Model (AAM) was for practising assembly of the heat shield and backcover of Beagle 2 and the packing of the gas-bags and parachutes.

The Electrical Test Model (ETM) allowed engineers to debug all the electrical functions, test the telecommunication system and operating software. Mechanical engineers used a Mass Stiffness Model (MSM) which had the rigidity properties, but not the actual shape of the Beagle 2 probe, to attach to the Mars Express orbiter for vibration, acoustic and balance tests; the MSM2 had a working spin-up and eject mechanism, the device which sends Beagle 2 spinning into the martian atmosphere.

The Volume Interface Model (VIM) was the same size and shape of the Beagle 2 probe, without its mass or mechanical properties, to ensure MEx and the lander will fit into the space on top of the Soyuz-Fregat rocket.

Any model that has exactly the same components as the ultimate flight model all connected together is known as a Qualification Model (QM). These were used to demonstrate that the flight model is qualified to survive the rigours of the mission. And finally, the last version, the flight model fondly known as the FM, gets to go to Mars.

A wild scheme

When Charles Darwin asked for his father's permission to join HMS *Beagle* on a voyage to South America, to act as the ship's naturalist and companion to her commander Robert FitzRoy, Robert Darwin was opposed to the idea. Having seen his son abandon medicine at the University of Edinburgh, as soon as surgery was introduced to the curriculum, and now seeking to avoid entering the church, he gave Charles a

Darwin's list of objections to his going on the Beagle voyage as raised by his father by permission of the Syndics of Cambridge University Library.

(1) Disreputable to my character as a Clergyman hereafter

(2) A wild scheme

(3) That they must have offered to many others before me, the place of Naturalist

(4) And from its not being accepted there must be some serious objection to the vessel or expedition

(5) That I should never settle down to a steady life hereafter

(6) That my accomodations would be most uncomfortable

(7) That you should consider it as again changing my profession

(8) That it would be a useless undertaking

variety of reasons why he was against the plan. Not least of these was "it's a wild scheme, no good will come of it"!

As Darwin senior pointed out, Charles was not the original choice for the expedition. A front runner was John Stevens Henslow, Darwin's mentor at Cambridge, who declined but suggested his young protégé. FitzRoy was also believed to have favoured a man called Henry Chester with whom he was already acquainted.

Robert Darwin by English School. Courtesy Darwin College, Cambridge and the Darwin Heirloom Trust.

Not wanted on voyage

One might think that the idea of adding a lander to Mars Express to search for life would be instantly acclaimed. But like Darwin's request, when it was first proposed, the idea of Britain providing the landing spacecraft and the instruments to look for life on Mars was dismissed out of hand. It's been done before and the results were negative; it will jeopardise the Mars Express orbiter; there's not enough payload mass for experiments; it's a gimmick, a stunt; it will wreck plans for other space missions; NASA are going to do it when they bring back samples; the Soviets have tried to land and failed, why should Britain think it can succeed? These were some of the many reasons against Beagle 2 given by its early opponents, which initially included ESA and the UK Government.

Oh and yes, there was already another candidate for landing on Mars, a proposal to build a network of meteorological and seismological stations proposed by a group of French and Finnish scientists. It had been offered for a number of possible missions to Mars. Now that ESA was having its own mission, the Netlander team were confident of selection. ESA had every cause to be worried about British commitment. In the past the British did not have a good record of supporting ESA's space missions.

"Mars or bust! No thanks – we're British
Whilst space bureaucrats in Moscow and Washington argue furiously about the best way to land people on Mars, their equivalents in London petition Whitehall for the cost of the daily tea-trolley they don't yet have."
The *Sunday Times* August 9th 1987

Throughout the build of Beagle 2 the project has been reminded constantly that first and foremost Mars Express is an orbiter mission. If Beagle 2 was not ready on time then it would not fly to Mars.

Even in early 2003, Britain going to Mars still seemed unreal.
"It [Beagle 2] all sounds incredible – and it is. A British space probe! That could discover life on Mars! Britain doesn't do space, do we? Why haven't we heard of this wondrous thing? Sorry to be annoyingly unscathing about this but it deserves our support."
Private Eye January 2003.

If you can find any man of common sense

Robert Darwin senior was obviously leaving the door open more than a crack when he gave the biggest possible hint that if Charles's uncle, Josiah Wedgwood II (son of the famous potter), thought the voyage of HMS *Beagle* was a good idea he would be willing to relent. It was Uncle Jos's

Josiah Wedgwood II by William Owen, courtesy of the Wedgwood Museum Trust, Barlaston, Staffordshire (England).

strategy for Darwin to list all his father's objections so that each of them could be challenged or refuted. Armed with the case for the opposition, Josiah Wedgwood simply sat down and composed a letter.

Robert Darwin's antagonism was less to do with fear for his son than concern for his own reputation. The philosophical solution Uncle Jos put forward was of the ilk it might 'make a man of Charles'. It was a case of a few years as a naturalist on a boat in South American waters cannot do any more harm and might do some good.

Within hours of sending his message, Uncle Jos was able to gather up a not-too-despondent Charles from his favourite pastime of blasting away at the local partridges on the first day of the shooting season, visit Dr Robert and receive a father's blessing and "all the assistance in his power".

Darwin's attempt to reconcile his father was along the lines 'look on the bright side, I'll find it difficult to spend your money stuck on a boat at the end of the Earth'. To which his father retorted "but they all tell me you are very clever".

The Wedgwood Family portrait by George Stubbs, 1780. Uncle Jos is seated on the pony (4th from right) whilst Charles Darwin's mother dominates the scene on the horse in the centre. Image by courtesy of the Wedgwood Museum Trust, Barlaston, Staffordshire (England).

A band of supporters

Darwin was very lucky that his father asked him to find only "one man of common sense" who would say that a voyage to South America was the chance of a lifetime. The Beagle 2 project's task was infinitely more demanding. It needed a million voices pressing the case, organisations and individuals, who by their very existence would convince the authorities to contribute to the budget, or persuade sponsors to back their judgement with contributions, advertising revenue or donations. So it embarked on a publicity campaign. Public opinion was to be Beagle 2's strongest card.

BBC News presenter Emma Simpson: "with Blur's help this will be the hippest ever venture to another planet."

For Beagle 2 the turning point was not the intervention of one man but a group, or to be more correct a band, the rock band Blur and with them they brought Damien Hirst, the *enfant terrible* contemporary artist, best known for sawing cows in half and pickling sharks.

And money, Beagle 2 had none, it relied on the fact that a number of people agreed to work on the project at their own cost. The lobbying to get the project technically selected and funded was not over in a morning, it took several years.

The author, Alex James, Damien Hirst and Dave Rowntree at the Royal Society's New Frontiers in Science Exhibition, 1999.

Alex James tells Lord Sainsbury "we've had Britpop and Britart, now its time for Britspace", at the Farnborough Airshow 2000.

Not just a publicity stunt

Alex James, the bassist, and Dave Rowntree, drummer, of Blur joined the Beagle 2 team to help with PR in the summer of 1998. Since then they have attended numerous promotional events on behalf of the project. The band's unique contribution is writing the Beagle 2 signature tune, a call sign to be beamed back from Mars to announce the lander has arrived.

"...kind of like a musical cave painting, a ponderous, clear tune." Alex James quoted on the *New Musical Express* web site.

The refrain was composed in 1999 and released on a CD along with *No Distance Left To Run*, the third single from Blur's platinum album '13'.

The musical signal actually comprises a series of notes, sounding rather like a mobile ring tone. The software destined to go onboard the lander was coded by entering a single bit for each frequency and the messages from Mars will be decoded to reveal the Blur composition.

Alex James explained that the notes which make up the call sign are loosely based on a Fibonacci series: "A mathematical sequence of notes with a little bit of artistic interpretation" adding "you have to be quite tasteful when you're going into space – it's a massive nature reserve".

Transferring the Beagle 2 call sign to the lander's software in the recording studio with Alex, Damon Albarn and Ben Hillier.

Spots on Mars

Damien Hirst has produced one of his trademark spot paintings to act as the calibration target for Beagle 2's cameras and spectrometers. The target is fitted to a place on one of the ribs of the lander so that the instruments can focus on it to check their readings. Several of the instruments need to be calibrated once on the surface of Mars. The spot painting has all the necessary ingredients: a selection of colours, different compositions and can be made small and compact.

"I'm sure there'll be a great demand for my work out there – they'll love me!"
Damien Hirst

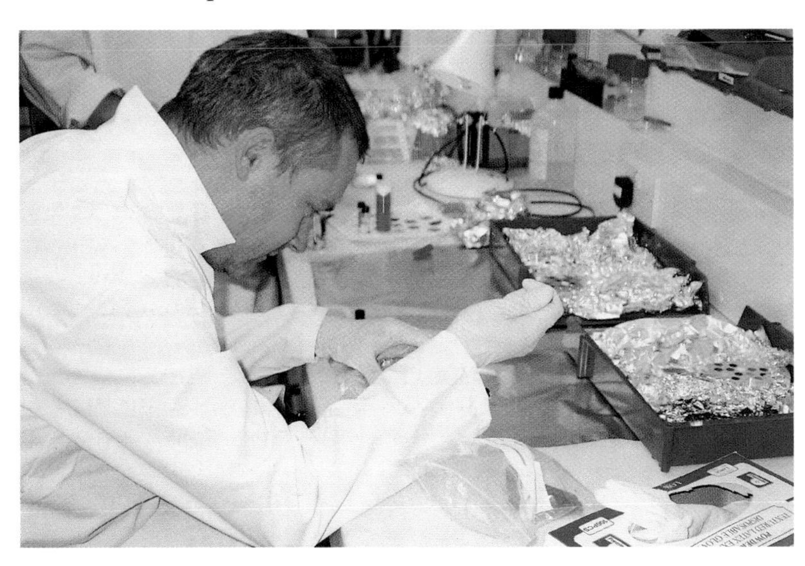

Damien Hirst in the laboratory working on the calibration target.

With the finished Beagle 2 spot painting, titled Mars, on show at the White Cube Gallery in London.

Although we humans see the chart only in terms of different colours, the instruments will detect other characteristics. The chemical signatures of the pigments reflect the minerals included in them. The target consists of an aluminium plate with indentations cut in the metal to accept the characterised and coloured spots. The whole piece measures less than eight by eight centimetres and weighs only twenty-six grams. The pigments are suspended in a type of clear adhesive which has been proved to be suitable for the space missions. This means, in the jargon of space engineers, that it is 'space qualified'. It will withstand extremes of temperature, vibrational shaking, the shock of impact and does not evaporate under the vacuum of space which could result in residues being deposited on other sensitive parts of the spacecraft.

The pigments themselves are rather appropriate considering the destination. Red and yellow ochre, naturally occurring iron-containing minerals, have been used as colouring agents for many centuries. At the onset of the Industrial Revolution, artificial iron pigments were made as by-products of some chemical processes, though alchemists were producing them earlier. Although the natural ochres were not in short supply, the man-made ones offered a more consistent material. The first identified synthetic iron pigment was *Mars Yellow*, a late eighteenth century translation of the latin *crocus martis*, *crocus* being derived from the yellow of the saffron crocus and *martis* from the word for iron.

Mars Yellow, when heated, becomes *Mars Red*. Gradually

A cartoonist's comment on the art work and its mounting. Published in Private Eye, with thanks to Andrew Birch, copyright Pressdram Limited, 2002.

a whole range of synthetic materials emerged from yellow through orange and red to violet, brown and finally black – called the *Mars Colours*. The change in colour is due to the relative amounts of the different forms of iron; the black pigment is the most highly oxidised.

The calibration target incorporates nine synthetic Mars iron oxides of different shades of yellow, red, orange, etc. These will allow scientists to match accurately the colour of the martian landscape. One of the samples is a synthetic pigment, actually produced at a time when Beagle 2's namesake HMS *Beagle* was circumnavigating the world. Another colour in the painting is called *Green Earth* a mixture of different oxidation states of iron as a hydrated silicate.

The *Mars Colours* satisfy many of the needs of the Beagle 2 instrument package but not all of them. A white spot was needed for contrast against black; titanium oxide is a white pigment and an important geological constituent. Azurite, a copper carbonate mineral, forms the blue spot. Other pigments contain cobalt and manganese, geologically interesting trace elements.

As well as playing a vital scientific role, the contribution from Damien Hirst allows the project to include an element of art without in any way unnecessarily using up the mass and power budget of the spacecraft, parameters which cannot be exceeded.

A sketch by Syms Covington of the Beagle *at the entrance to the Santa Cruz River, reproduced courtesy of the Mitchell Library, State Library of New South Wales.*

Musical manservant

On May 22nd 1833 Captain FitzRoy appointed one of the ship's complement to be Darwin's personal servant. Seventeen-year-old Syms Covington, "Fiddler and boy to the poop-cabin" as he was described, was presumably a source of music for the crew of the *Beagle*. Covington served Darwin until 1839 when he emigrated to Australia. He was obviously multi-talented: he painted pictures during HMS *Beagle*'s voyage as well. He was Blur and Damien Hirst rolled into one.

Money troubles

During the voyage of the third *Beagle*, Darwin was forever writing home to say he was being as careful as possible with his father's money. Darwin Senior had shelled out two hundred pounds to settle his son's debts at Cambridge and another six hundred pounds, equivalent to two years at the University, to equip him for and finance the voyage round the world. But Darwin was never really strapped for cash.

Robert FitzRoy on the other hand had more than his fair share of difficulties concerning funds for the expedition but at least Darwin was not a financial burden on him. FitzRoy managed to persuade the Admiralty to allow him to include the naturalist amongst the official ship's crew, although for the benefits of dining with the Captain, Darwin had to pay fifty pounds per year.

Medal struck to commemorate the first voyage of HMS Beagle *with the original* Adventure *during 1826–1827. Courtesy National Maritime Museum, London.*

Surprisingly, however, the instrument maker George Stebbings came entirely at FitzRoy's expense, despite 'The Orders', the Admiralty instructions to the Captain, concerning the chronometers. These required a great deal of effort to maintain in order to complete a worldwide network of longitude reference points.

The real disagreement between FitzRoy and the Admirality was over a companion ship for *Beagle*. Without one the Captain felt he could not possibly achieve all the tasks he had been set. HMS *Beagle* was just too small to carry everyone and the baggage. And besides, a tiny ship alone in the wastes of the southern ocean was not safe. After one particular violent storm FitzRoy determined to acquire a tender. Using his own money, he purchased and refitted the *Unicorn*, which he renamed the *Adventure* as a reminder of *Beagle*'s first voyage. When he wrote to Captain Beaufort at the Admiralty to seek permission to make the purchase he was full of optimism that their lordships would understand.

The reply FitzRoy received was very negative: the Navy could not support these extravagances. FitzRoy was forced to sell his *Adventure*, dispense with the crew of twenty he had recruited, and recall John Wickham who he had installed in command. The rejection of his plans contributed to the fits of depression the Captain suffered.

FitzRoy wrote that selling the second *Adventure* cost him dear, the price he got being much less than the sum he paid. Darwin's comment on the affair was that the Admiralty treated FitzRoy "meanly" because he was a Tory and the Whigs where in power (Darwin himself came from a Whig family).

No ship ever put to sea without Treasury support

The biggest problem that Beagle 2 had to overcome was raising the money. Normally space missions are years in gestation so when they are finally approved Government Agencies have money put aside in their budget to fund them. Mars Express was not like this; the opportunity for a 2003 launch was so good that ESA just had to take it with only a six year lead time which is incredibly short for assembling a space mission.

It was not a problem for most people who wanted to be involved; they had spare instruments from Mars 96 and

The cartoonists enjoyed Beagle 2's money problems. 'Who will pay ?' by Peter Shrank, first published in the Times Higher Education Supplement.

therefore needed only moderate amounts of funding to get ready to fly on the replacement orbiter.

The attraction of looking for life on Mars however was irresistible and the Beagle 2 team began to work without any visible means of financial support. Like FitzRoy they agreed that they would initially pay for Beagle 2 out of their own resources. A new term for it was coined – 'Marslighting'.

The difference between HMS *Beagle* and Beagle 2 was that the powers that be eventually became convinced that they should help finance the great adventure of landing on Mars. By the summer of 1999 the technical achievements and the PR campaign had demonstrated to the UK Government that there was something in this Beagle 2 business. So an announcement that ESA had agreed to take the tiny spacecraft released a tranche of Government money, about £8M, to the industrial partners for the lander itself and a contribution to the university groups to build the instruments.

Of course for the whole project this was not nearly enough; the initial estimate in 1997 was that it would cost £25M. Realising more would be needed to get to the launch pad, other sources were approached. After a review of the project by an independent NASA panel, which stated that it was not advisable for Beagle 2 to carry on without a detailed test programme to identify areas of risk, ESA were persuaded to contribute. But even this was not enough, so the balance of what was thought necessary was borrowed on the strength of a business plan which suggested that Beagle 2 could attract sponsorship and advertising revenues.

The Beagle - Temperament - cheerful and friendly but with a tendency to wilfulness.

Beagle 2 and the City bankers, by Phil Emms.

So just as Darwin and FitzRoy set off from Devonport in 1831, full of optimism in the expectation that they were on a three year journey, in 2000 the Beagle team embarked on a path which could happily take the spacecraft to Mars in 2003. Whether the team will leave behind them debts like Darwin is anybody's guess. One of Darwin's friends unkindly said that Charles only went on the voyage of HMS *Beagle* to avoid paying his creditors or as she put it "a pretty drastic way of avoiding one's tailors". Maybe it had been a necessity: Henslow was still sending bills to Robert Darwin two years after the ship sailed.

Beagle 2 was never going to be big enough to carry one member of the space project team, so stowing away to avoid repaying the loan will not be an option.

Lobbying at the Houses of Parliament. Carrying the Beagle 2 model, the author, John Underwood (Martin Baker Aircraft Company) and Mike Rickett and John Thatcher (Astrium).

As much of an Englishman

Darwin's trip on HMS *Beagle* was the second of three voyages of discovery that the ship made. The third voyage concentrated on expanding commerce with Australia before anyone else got there, particularly the French, whose ships were often better supported.

On September 21st 1837, when *Beagle* was anchored in Simon's Bay, just around the Cape of Good Hope *en route* to Australia, for a trip which would involve twice circumnavigating the new continent, she was joined by another exploration ship, the 36 gun French frigate *Heroine*. Pleasantries were exchanged between the two crews and their captains, HMS *Beagle* now being commanded by John Wickham, who had succeeded FitzRoy. Destinations were discussed and although the French said they were bound for Tahiti, the British suspected their true destination was Van Diemen's Land (Tasmania) or New Zealand.

Artist's reconstruction of the layout of HMS Beagle *showing where supplies were stowed.*

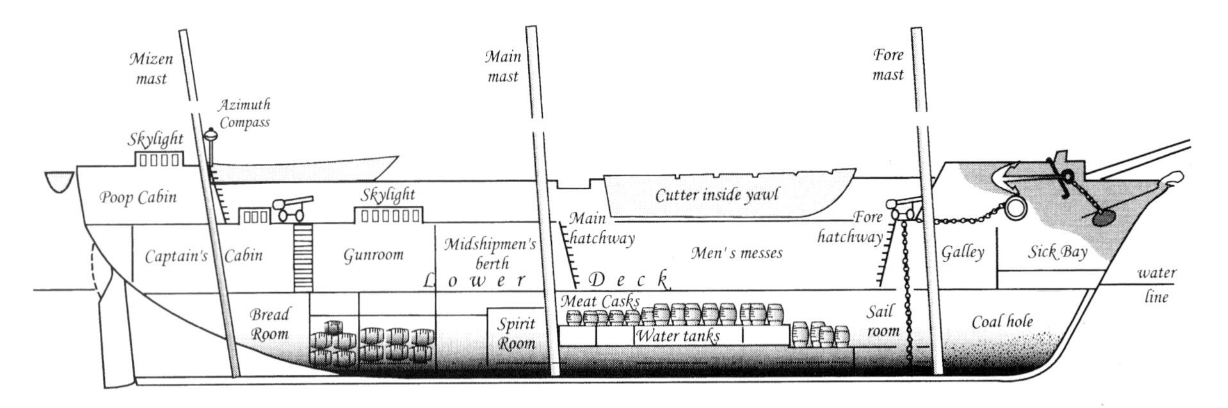

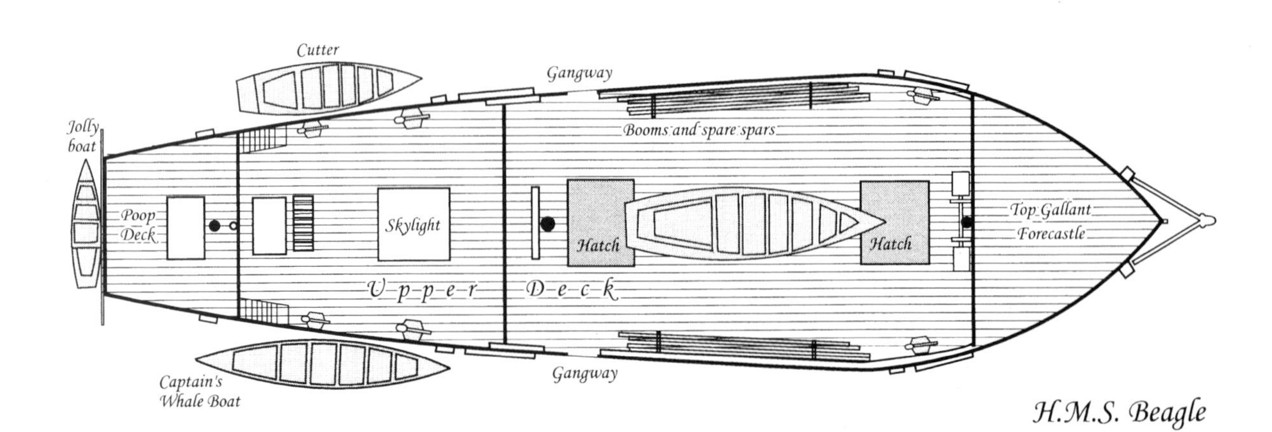

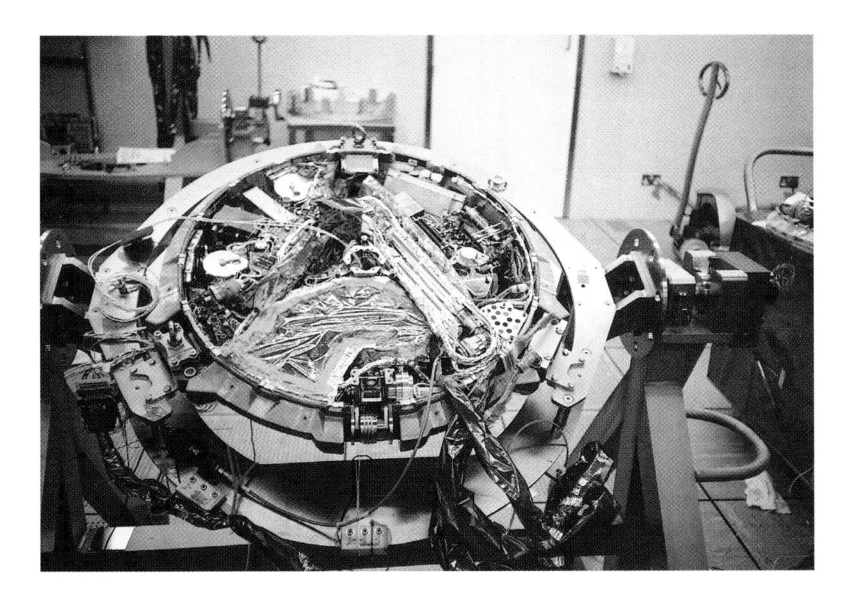

The Beagle 2 lander just prior to attaching the lid. Damien Hirst's spot painting can be seen at the bottom right.

Wickham, with *Beagle* loaded down to the gunwales, with every available space taken by the stores for an eight month voyage, wrote to his superior, (now Sir Francis) Beaufort: "what a difference between the French and English surveying ships ...the one with more than twelve months provisions all stowed out of sight and the other (HMS *Beagle*) with scarcely a passage fore and aft the deck, but I am as much of an Englishman to think that we look the more fit to work."

There is no doubt that Beagle 2 is tightly packed, it was described during an ESA review as "the most ambitious science payload mass to that of the system support mass of any spacecraft ever attempted".

Then there was one

Like the *Heroine*, the alternative proposal to Beagle 2 for Mars Express, Netlander, championed by French scientists, fell by the wayside as a result of being too big for the purpose. Initially it was believed that Mars Express would have ample capacity for landers. The Netlander team who had been studying the possibility for landing on Mars since the beginning of the 1990s was adamant that a network needed to be four stations. This meant they required something over 200 kilograms of the Mars Express mass budget. The Beagle 2 team offered to share the resources with Netlander by building a large lander to carry the experiments needed to search for life plus a subset of Netlander instruments. It could act as one of the stations of the network. Meanwhile the Netlander team could provide two landers to complete a triangle of three stations.

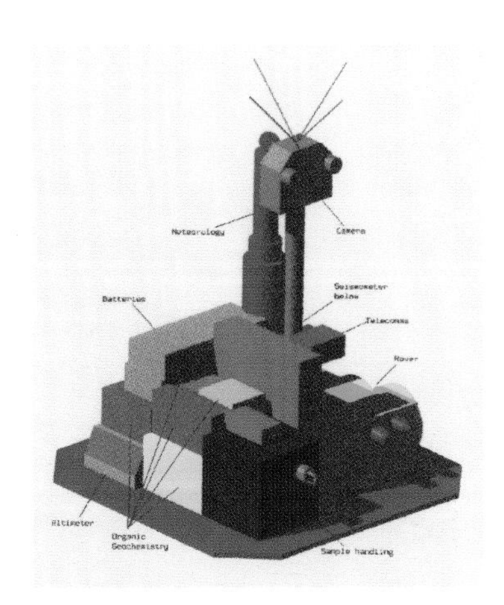

A concept picture of how Beagle 2 might have looked if enough mass had been available.

The idea never got very far because soon ESA announced it could only spare sixty kilograms for a lander or landers. Since no one had proposed to build even a single lander this small, Mars Express would be an orbiter mission only. Netlanders were immediately out but ESA reckoned without the resourcefulness of the Beagle 2 team.

Needs must

It was all very well FitzRoy deciding to take Darwin along on HMS *Beagle* but where was he to put him? The tiny ship had only enough room for its compliment of men so somebody had to be left behind. The obvious first choice for redundancy was the Padre – FitzRoy decided he could double in this role. This was a bit of an understatement as he was already God on board according to Navy rules and regulations. He could flog the crew (he actually punished four men the first day out of Devonport) and, as was the case with the fifth *Beagle*, he could put a man to death if deemed appropriate (he never had the need).

Another problem was that Darwin was a tall man, just slightly short of six feet. When he inspected the accommodation he was assigned, he found his sleeping space far too short. So what did he do? Every night for five years he removed the drawer from the bureau to give himself somewhere to put his feet.

Darwin shared the poop cabin, a space eleven feet by ten feet, with the Beagle's surveyor, John Lort Stokes and

hasty word of, or to, anyone. You will therefore readily understand how this, combined with the admiration of his energy and ability, led to our giving him the name of "The dear old Philosopher" – Stokes has given a good idea of the way in which he struggled against sea sickness – but he has not said that the narrow space at end of the chart table was his only accomodation for working, dressing, and sleeping. The hammock being left hanging over his head by day when the sea was at all rough, that he might lay in it with a book in his hand when he could not any longer sit at the table. His only stowage for clothes being several small drawers in the corner

The poop cabin showing Darwin's hammock and the drawers as Bartholomew Sulivan remembered it, sketched in a letter to T. Huxley, by permission of the Syndics of Cambridge University Library.

midshipman Philip Gidley-King during the day. At night Stokes was forced outside under the companion way; King slept with the crew. When Darwin was ill, as he frequently was during his time at sea, he lay in his hammock suspended above his compatriots working at the chart table.

FitzRoy created a separate tiny cabin for himself which Darwin was entitled to share but he had to vacate if the Captain requested his privacy.

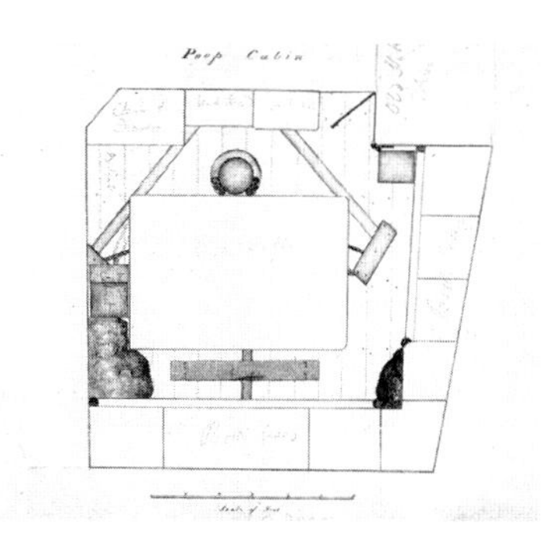

A sketch of the poop cabin by Philip Gidley-King. By permission of the Syndics of Cambridge University Library.

Puppy version

This was definitely a case of *déjà vu*. Could the Beagle 2 team improvise like Darwin? The simple answer was yes. The size of Beagle 2 as originally proposed to ESA was 108 kilograms, being driven by the need to have a vehicle to collect samples. When the US Pathfinder mission landed with its rover Sojourner in 1997 it had been an amazing PR success travelling from rock to rock. The Beagle 2 team wanted a means to move away from the lander to increase the choice of sampling sites, but to reduce the size of the lander Beagle 2's rover had to go.

The requirement to have mobility was achieved when it became apparent that the device which Beagle 2 was to use to collect subsurface samples of Mars, the mole, could also crawl across the surface and burrow under rocks. Instead of a vehicle, the lander would then need a mechanical arm to deploy and retrieve the mole, but with this it could reach out and recover pieces from rocks close by or press instruments

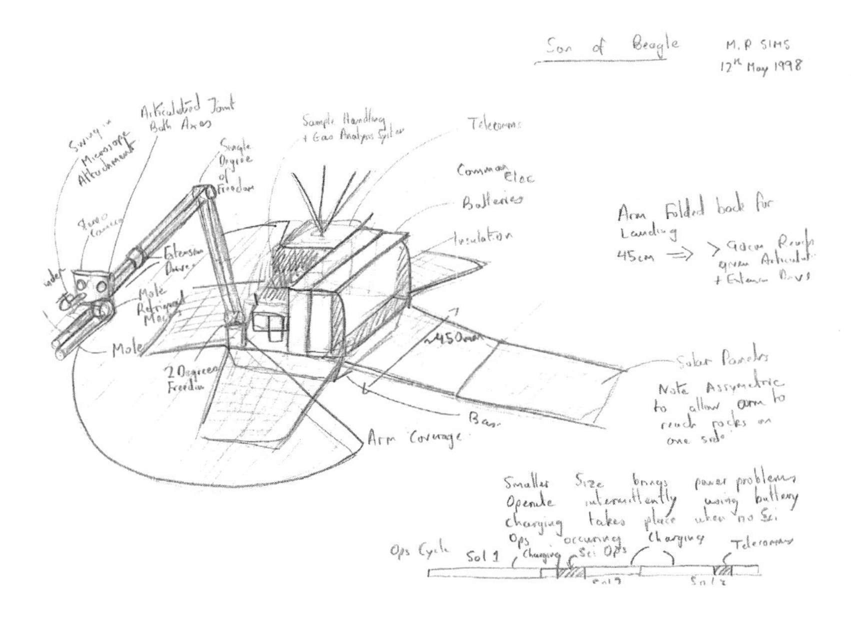

Concept sketch of the scaled down lander, thanks to Mark Sims for permission to reproduce.

up against them. The mole soon acquired a spacey name with a canine connection, PLUTO – the Planetary Underground Tool.

The real trouble with a vehicle was the space needed to accommodate it, so removing the rover allowed Beagle 2 to be shrunk in three dimensions, which reduced the lander mass significantly, and everything to deliver the lander to the surface could be scaled down accordingly.

Within a short while of being told by ESA there was no lander of about sixty kilograms, the Beagle 2 team had invented one without sacrificing a single bit of science except the Netlander's seismometer, which was superfluous without the analogous devices making up the other portions of the Network. Beagle 2 (the puppy version) was simply a miniature of the much bigger original. The robotic arm was stowed for launch by wrapping itself around the lander body inside fold-out solar panels. It was almost as though it was giving itself a big comforting cuddle to reassure the Beagle 2 team it was still in business.

The eventual mass budget allocated to the Beagle 2 probe was sixty-eight kilograms. The additional eight kilograms

The Ship that Sailed to Mars by William Timlin. "The Old Man made with loving care a big working drawing of the Ship and everything was marked down on it. It nearly came about that the drawing, of which they were all very proud, was spoilt by one of the Fairies, who would persist in colouring it blue in the wrong place."
Note the man at the bottom with the scales weighing all the components.

came when ESA struck a deal with Starsem, the company that supplied the rocket, to increase the launch mass of Mars Express. Another five kilograms needed by Beagle 2 remain on the orbiter as the the spin-up and eject mechanism, together with electrical circuits to fracture bolts holding the lander in place and cabling to provide the probe with power and charge its batteries for the voyage to Mars.

About half of the 68 kilogram probe is made up of the entry, descent and landing system, thirty-five kilograms are required to make sure the lander reaches the surface intact. More systems mass is taken up by the structure, the solar arrays, the batteries, and insulation. The scientific instruments account for just over nine kilograms whilst the robotic arm, an essential item for recovering samples, is slightly more than two kilograms.

To the credit of the Beagle 2 team, when the probe was weighed just before delivery to Mars Express, it was within a hundred grams of the mass estimated more than two years earlier.

Deposed

Not everyone was delighted with Darwin's recruitment to HMS *Beagle*. The job of naturalist usually fell to the ship's surgeon, the most scientifically educated man on board. When the surgeon was not looking after the well-being of the crew he could collect specimens as decreed by the ship's orders.

The puppy model on show at Farnborough Airshow in 1998 under the headline "Putting Britain on the map – of Mars".

The surgeon assigned to HMS *Beagle* was Robert McCormick. In a letter home Darwin wrote "I take the opportunity [of writing] of McCormick returning to England, being invalided (that is being disagreeable to the Captain) – he is no loss". McCormick had begun grousing even before he left Devonport, arguing about what colour his cabin should be, French grey or lead white. "The surgeon is an ass" was Darwin's comment.

In his memoirs published in 1884 (two years after Darwin died) McCormick was still at it, he said "having found myself in a false position on board [*Beagle*], and very much disappointed in my expectations of carrying out my natural history pursuits, every obstacle being placed in the way of my getting on shore and making collections, I got permission from the Admiral in Command of the station to be superseded and allowed a passage home". He was so cross he could not bring himself to mention Darwin by name after fifty-two years.

When Beagle 2 was accepted for the Mars Express mission, the Netlander team were probably not too pleased about missing yet another chance of going to Mars either.

From a Brig to a Barque

Although HMS *Beagle* was designed and built as a brig she never put to sea with two masts. Five years after launch she returned to Woolwich dockyard to be re-rigged as a barque, that is with three masts and a different arrangement of sails.

Slinging the monkey, a game played by sailors of HMS Beagle *on Christmas Day 1833, painted by Conrad Martens. The picture is annotated 'The main mast of* Beagle *a little further aft, mizzen mast to rake more.' The changes were initialled FR by Robert FitzRoy, by permission of the Syndics of Cambridge University Library.*

The modification gave her even greater manoeuvrability and meant she could be handled by a skeleton crew, vital because as a survey vessel there would be times when a large contingent would be ashore on mapping sorties.

Total refit

Having been changed in design before she set sail, HMS *Beagle* was decommissioned and re-fitted at least twice more in an as-good-as-new form, first and most famously when Captain FitzRoy prepared her for her second voyage: Darwin's great adventure. For this some new timbers were required with mahogany being chosen for strength.

FitzRoy had already experienced the 'joys' of a voyage back to England from South America on board the tiny HMS *Beagle* with its headroom of only fifty-three inches. His first modification to the ship before setting out on the second voyage was to raise the deck twelve inches in the forward areas and eight inches aft, to avoid the hazard of banging his head. Having dealt with areas of the ship above the water he turned his attention to below. He ordered the bottom to be clad with an extra layer of two inches of pine planking, covered it with a layer of felt and on top of this a layer of copper. The copper was to prevent the wood from rotting and the iron nails from rusting.

The Captain equipped *Beagle* with a new rudder and a patent windlass in place of a capstan. The whole exercise added some fifteen tons to HMS *Beagle*'s displacement tonnage. Six small boats, two of them private property, were expressly built for HMS *Beagle*.

HMS Beagle *rigged as a barque. Courtesy National Maritime Museum, London.*

Artist's impression of Mr Frazer's stove, from the patent details lodged in 1829.

FitzRoy also wrote in the *Narrative* of the *Beagle*'s voyage that amongst the improvements he made was to install "one of Frazer's stoves, with an oven attached, instead of a common galley fireplace" so that hot food could still be prepared even in the roughest seas.

These 'no expense spared' modifications cost £7,583 only £220 less than the original purchase price. The workmanship however was of such quality that after five years at sea, in the most inhospitable waters off South America, her next refit before the passage to Australia was achieved for just £2,384.

On the back of a beer mat

In the Autumn of 1998 the puppy version of Beagle 2 was selected by ESA and allowed to proceed as part of the Mars Express mission. No sooner than the proposal was accepted, the design was changed to make something much more appropriate to the challenges ahead. The corners of the pyramidal shape would be almost impossible to protect during the fall-out of the gas-filled bags needed to enclose the lander as it bounces to rest on the surface of Mars. There were too many mechanisms, motorised hinges, needed to unfold all the sections of the solar arrays which would generate the essential electrical power. Also there were too many ways up that the pyramidal Beagle 2 could land.

The idea for a round lander structure sketched on the back of a beer mat.
Courtesy Dr Mark Sims.

The final design of Beagle 2 turned out to be clam-shaped; other descriptions have likened it to a large pocket watch, a garden barbecue, and a bagel. Its diameter is almost the same as a standard bicycle wheel. Just like HMS *Beagle*, the first plans for the new Beagle spacecraft were sketched on

Photo-shoot with the full-size model of Beagle 2 showing the PAW and three solar arrays. A fourth was added later.

the back of something else – in the case of the Mars lander it was a beer mat.

Beagle 2 was modified several times after its metamorphosis into the barbecue-shaped design. For example, the pop-up telecommunication aerial, which originally looked like the coathanger favoured by Ford Cortina owners, was abandoned for a more conservative version incorporated into the lander lid. Likewise the number of circular fold out solar panels increased from three to four to obtain more power to charge Beagle 2's battery. Either of the above could be considered the equivalent of changing the number of masts on the naval predecessor.

A much more fundamental change occurred when the PAW was conceived. The PAW, or position-adjustable workbench, which allows instruments and sample-handling devices to be used sequentially without the need for picking up and putting down, greatly reduced the number of operations for the robotic arm. It removed the risk of failing to make good electrical contacts, or worst still losing a device altogether. After the initial jump to sixty-eight kilograms, most of the modifications to Beagle 2 were to help to keep down the mass not raise it.

Captain's cabin

Beagle 2's final design splits the spacecraft into three sectors, with three ribs to give structural strength. One sector is for the mass spectrometer and gas analysis package jokingly referred to as the Captain's cabin.

The lander shell showing the ribs of the three sectors construction.

Another sector houses the rest of the scientific instruments fixed onto the PAW at the end of robotic arm which folds across the lander.

The spacecraft's electronics module (ELM) which comprises the central computing, battery package, the electronics boards and telecommunications unit are contained in the third sector. The lid, connected by a simple, cable-driven hinge, has the telecommunications aerial and four arrays of solar cells, each with its own motor-driven hinge, sequentially folded into it.

Testing the motors needed to drive the joints of the robotic arm and the main hinge turned out to be quite a tricky task for Beagle 2's engineers. Because of the lower gravity on Mars the arm needs to move the equivalent of forty percent of the mass. Since there was no point in using more powerful, and

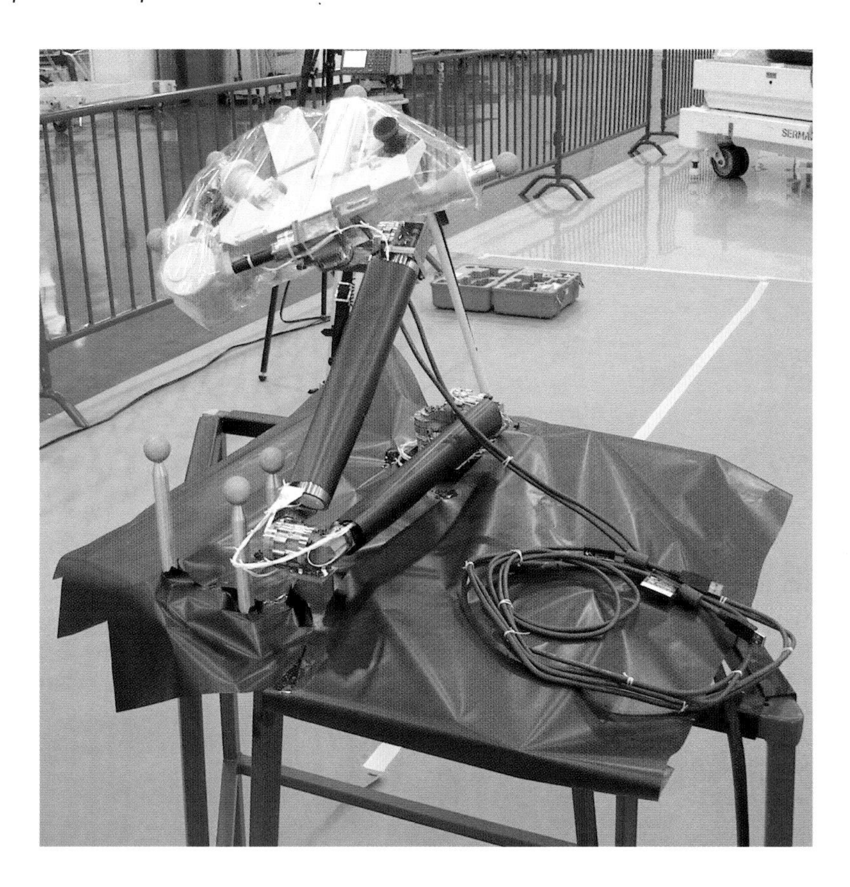

The robotic arm lifting a one third-weight PAW during a motor test at Astrium.

heavier, motors than needed on Mars, the tests on Earth were done with a lightweight PAW structure.

Racing ahead with the lander structure

Just as the hull of HMS *Beagle* was made up of layers of timber, felt and copper, the main structure of Beagle 2 is

The lander base under construction in the McLaren Composites factory. Some of the materials used are seen on rolls, right.

*The Ship that Sailed to Mars by William Timlin.
"The building of the Ship went very quickly. Everyone was so interested and beyond closing up one of their number in a cabin with no door, there were no mishaps, or serious problems of construction to face. The wood, of incredible lightness, was brought through a trap-door from the grove of a friendly gnome..."*

likewise a sandwich composition. It owes the design to the technological research and development associated with the sport of Formula 1 motor racing. The multilayer-skin, which shields the science instruments and other delicate parts of Beagle 2 from impact damage during landing, is similar to that which protects drivers travelling at 200 mph from the catastrophic effects of crashes.

The Beagle 2 design consists of an upper and lower composite sandwich which utilises four distinctive material types plus adhesives. The base, the lid and various fittings all have similar construction. The inner surface is a carbon-fibre laminate, bonded to an aluminium and fibre honeycomb core, which provides stiffness to the structure. Next comes a foam-like layer which acts to absorb energy and shock. Finally the outside layer includes Kevlar, best known for its role in making bullet-proof vests, as a shield against penetration by the rocks and boulders expected at the landing site.

Electric discharge

Thanks to Robert FitzRoy's awareness to scientific advance HMS *Beagle* carried lightning conductors. These were produced by inventor William Snow "Thunder and Lightning" Harris.

In his diary Darwin, who went to a public lecture by Harris before sailing on *Beagle*, described the inventor's system thus: "This plan consists of having plates of copper folding over each other, let in the masts and yards and so connected to the water beneath, the principle, from which these advantages arise, owes its utility, to the fact that the electric

The test set-up used by William Harris. Courtesy The Royal Society.

fluid is weakened by being transmitted over a large surface to such an extent that no effects are perceived even when the mast is struck by lightning".

For public demonstrations Harris used a machine for making electric discharges, a tub of water and a toy to play the part of a battleship. He would put gunpowder against the lightning conductor running down the ship's mast without any untoward effects. Darwin might not have got the principle of operation quite right but in practice Harris's method worked extremely well, and of course the copper bottom of *Beagle* helped.

Harris, despite being elected Fellow of the Royal Society for his invention, and FitzRoy's patronage, found it difficult to persuade the Navy to adopt his system for protecting ships at sea. In fact the Russians installed it on their vessels before it became common on British men-of-war. Ultimately the inventor got his just reward in the form of a knighthood.

Lightning conductors were fitted to all the masts, browsprit and flying jib booms of HMS *Beagle*. That they were a success there can be no doubt. The ship was struck whilst in Montevideo harbour. The officer on deck at the time, Bartholomew Sulivan reported:

"Having been on board HMS *Thetis* at Rio de Janeiro a few years since, when her foremast was entirely destroyed by lightning, my attention was always particularly directed to approaching electrical storms, and especially on the occasion now alluded to, as the storm was unusually severe. The flashes succeeded each other, the ship became apparently wrapt in a blaze of fire, accompanied by a simultaneous crash. One of the electrical clouds by which we were surrounded had burst on the vessel, and as the mainmast at the instant appeared to be a mass of fire, I felt certain that the lightning had passed down the conductor on that mast."

"Mr Rowlett, the purser, said he was sure the lightning had passed along the conductor, for at the moment of shock he heard a sound like rushing water along the beam [of which a main branch of the conductor passed]. Not the slightest ill consequence was experienced, and I cannot refrain from expressing my conviction that, but for the conductor, the results would have been serious." Sulivan was clearly given to understatement.

Conducting one's self properly

Static electricity is a perennial problem with space instruments both during assembly and when in operation. Technicians building instruments on Earth, in dry, filtered air in super-clean laboratories ground themselves before and during the time they are working on equipment.

In the latter stages of assembly when live pyrotechnic charges are installed great care has to be exercised because of the safety hazard. Those working on Beagle 2 wore double earth straps just to be sure.

Beagle 2 is gold-bottomed, entirely coated with a layer of gold, which will help to ensure that no static charge builds up to damage sensitive electronics by discharge when the lander is on Mars.

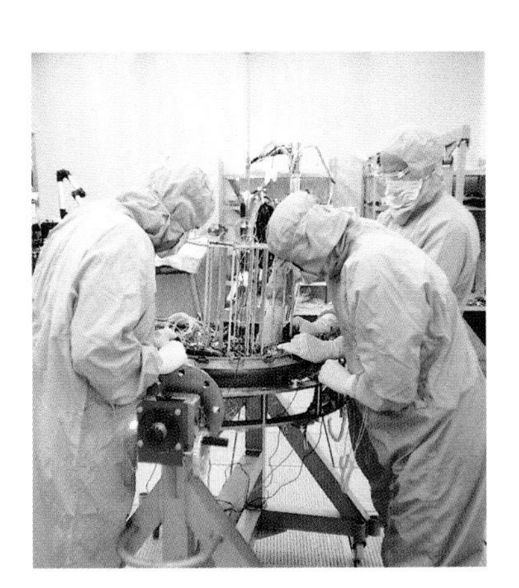

Technicians working in the Beagle 2 assembly facility showing the grounding straps.

Keeping warm

Keeping warm at night when the temperature plunges to −70°C is a priority for Beagle 2 since if the battery

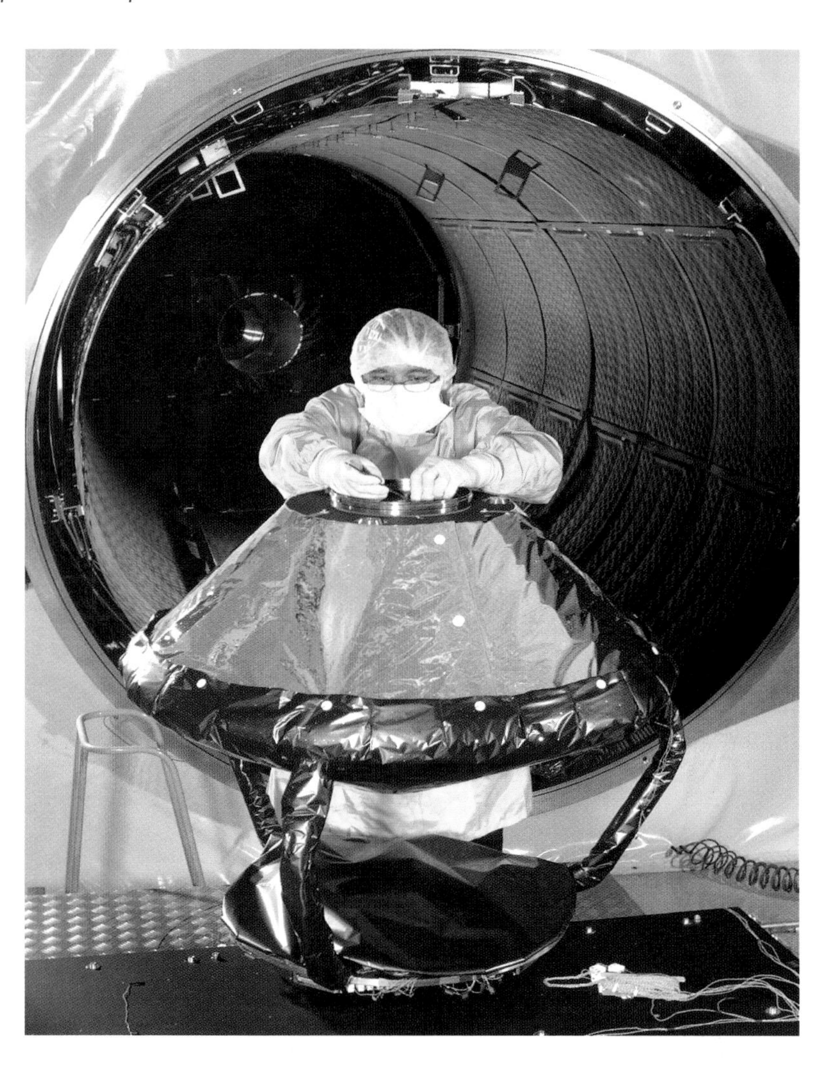

The Beagle 2 probe undergoing thermal tests at the Rutherford Appleton Laboratory.

temperature drops below –35°C its cells will freeze up and no power can be provided to the rest of the lander. So Beagle 2 will be protected by a multiple layer of insulation. It also has a solar radiation absorbing layer on its upper deck to store all the heat it can from the sun during the chilly martian day. The gas analysis package which generates heat during operation is contained inside the lander so that the electric power supplied is not wasted but used to keep the battery and other electronics warm.

Going for gold

Apart from protecting the spacecraft against electric discharges there is another good reason why Beagle 2 is coated with gold. Gold is a excellent 'solar absorber': it absorbs more radiative energy in the solar wavelength range than it emits at thermal wavelengths. This means that gold-coated surfaces tend to heat up in the sun.

The Beagle 2 lander with the solar absorber covering the electronics and GAP.

Spacecraft engineers refer to the absorption/emission ratio as the alpha to epsilon (alpha being the percentage of absorbed solar radiation and epsilon the percentage of emitted thermal radiation) ratio. Black surfaces absorb about the same proportion of energy as they emit therefore their alpha to epsilon ratio is one. Gold coatings have an alpha to epsilon of about ten, other materials for example aluminium surfaces are about three or four.

The upper surface (the solar absorber unit) of the lander is covered in gold-coated Kapton, which is a polyimide film used extensively on spacecraft. The Kapton is twenty-five

The closed-up lander. The shiny finish also helps to reduce radiative heat loss.

microns thick, whilst the gold coating measures about fifteen hundredths of a micron. For comparison, a human hair is about forty microns in diameter. Although the gold layer is thin, overall about a single gram of gold has been used on the outer surface of Beagle 2.

All the external surfaces of the lander, already coated with gold by vacuum deposition have a Kapton layer added for the same reasons as the upper deck. The vacuum-deposited layer inside minimises the heat lost from instruments by radiation through the walls.

Black cat under a tub

Oliver, one of the author's black cats. His whereabouts were carefully checked on Beagle 2 launch day.

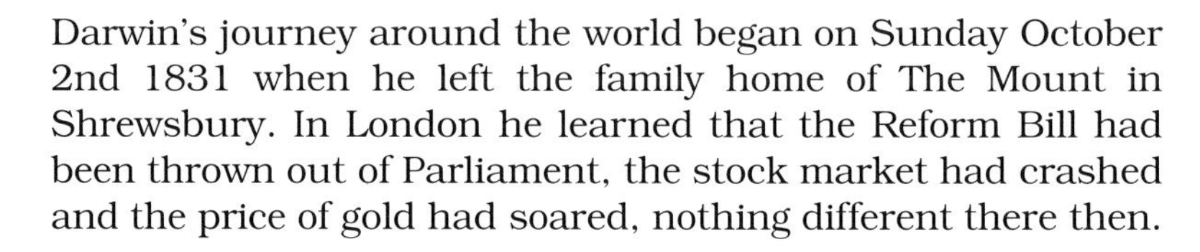

Darwin's journey around the world began on Sunday October 2nd 1831 when he left the family home of The Mount in Shrewsbury. In London he learned that the Reform Bill had been thrown out of Parliament, the stock market had crashed and the price of gold had soared, nothing different there then.

He had originally been told that HMS *Beagle* would sail in October. His luggage was despatched to Plymouth by sea, whilst he travelled overland to arrive on October 24th, ten days before a new departure date. In fact nothing happened for the whole of November, the 30th being set as the date when *Beagle* would be positively ready.

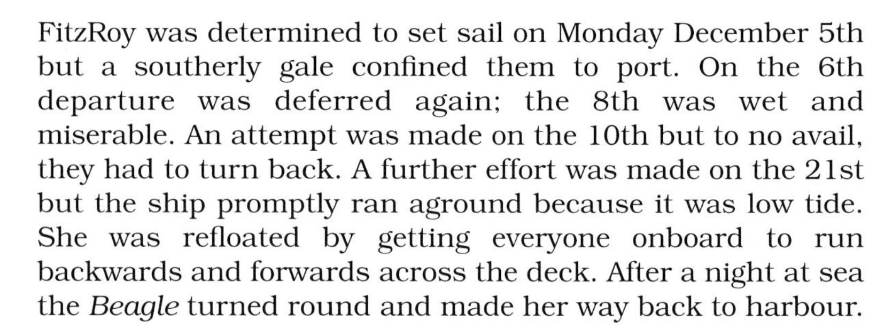

FitzRoy was determined to set sail on Monday December 5th but a southerly gale confined them to port. On the 6th departure was deferred again; the 8th was wet and miserable. An attempt was made on the 10th but to no avail, they had to turn back. A further effort was made on the 21st but the ship promptly ran aground because it was low tide. She was refloated by getting everyone onboard to run backwards and forwards across the deck. After a night at sea the *Beagle* turned round and made her way back to harbour.

On December 27th, whilst the *Beagle* tacked out of the port yet again, Darwin, FitzRoy and Sulivan dined aboard the local naval commissioner's yacht – to celebrate the start of a successful voyage. They joined their vessel outside the breakwater on a journey around the world which would take them at least forty thousand miles. Darwin was away from home for five years to the day since he arrived back in Shrewsbury in time for breakfast on October 2nd 1836.

When a ship experiences difficulty in leaving port superstitious sailors believe that somewhere somebody on shore is "keeping a black cat under a tub". Darwin wrote this piece of folklore in his diary on December 20th 1831.

Keeping to the launch schedule

Beagle 2 took an equally convoluted route before it left for Mars and the timing was just as imprecise.

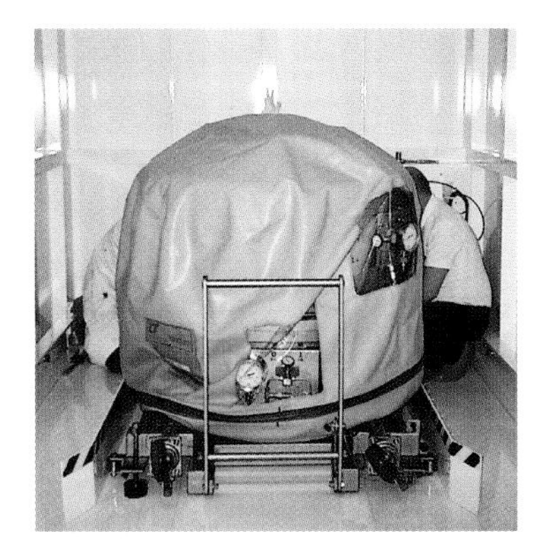

The lander was packed with parachutes and gas-bags into the probe which was then put into the spacecraft-like pressurised container inside the transport box.

The Beagle 2 project had aspirations of completing the assembly of the spacecraft before the end of 2002. This was soon changed to a more realistic January 15th 2003. Then, by agreement with ESA, to the 31st. Near the end of January the project was informed that the probe would not be needed for a fit check for another week, so it commenced its journey to join Mars Express in Toulouse on February 9th. But first it had to undergo vibration testing as a complete entity. During this process the probe was filled with gas at too high a pressure so it was returned to the assembly facility to be checked out and given the all clear. Its journey by land and sea in a specially built transport container to be mated to MEx in Toulouse finally commenced on February 23th.

But that was only the start of the journey. Both Beagle 2 and the orbiter travelled by air on to Baikonur in Kazakhstan via Moscow on March 19th. The Baikonur Cosmodrome is not in Baikonur at all but more than a hundred kilometres away, the misnomer being a relict of misinformation from the days of the cold war. The last part of the journey to the launch pad was on top of the Soyuz-Fregat on a slow-moving railway.

Large lorry, small box. The Beagle 2 convoy leaves The Open University where it had been assembled.

At first it was thought the launch date for Mars Express would be in the period June 1st to June 11th. That got brought forward to May 23rd when the capacity for more fuel was added. The firing of the rocket was scheduled for 7.40pm BST, 0.40am in Baikonur. As a consequence of having more fuel, the arrival date on Mars shifted from 26th December to 24th then to the 23rd, when touch down was anticipated for 1.15 in the morning. Delays to Mars Express subsequently pushed the launch date back again to June. Blast off on June 6th would ensure arrival on December 27th at 2.18 in the morning. When some time was made up, the launch date was changed again to June 2nd so that arrival moved to Christmas Day.

If someone was keeping a black cat under a tub, the launch could have been delayed until the end of June but Beagle had to be off by June 28th at the latest or the wait would be for 26 months. The final decision to launch the rocket was taken two and a half hours before the event.

Then there were two: Mars Express and Beagle 2 are loaded onto the Antonov transport plane at Toulouse for the flight to Baikonur.

The Beagle voyages

It took FitzRoy and Darwin five years to circumnavigate the World.

HMS *Beagle* herself made three voyages, to South America and back from May 26th 1826 to October 14th 1830; around the world from December 27th 1831 to October 26th 1836 and finally a return trip to Australia from July 5th 1837 to September 30th 1843.

Essential to successful navigation was being able to determine the speed of the ship. In the early days a knotted line was thrown over the side for a fixed duration, hence the term "knot" as a unit for speed at sea. By 1831, a brass propeller-like gauge and a counter had been substituted. By dragging it for a fixed time the speed of the ship could be read from dials attached to the line. This would be done a number of times every day.

The journey to Mars

The travel time to Mars will be seven months. After Beagle 2 took off aboard the Soyuz-Fregat it first circled the Earth, taking about 90 minutes to do so.

It is difficult to provide information about the speed at which spacecraft progress. When a space voyage begins it is the speed relative to Earth which is important but when it arrives then the velocity compared to the destination needs to be considered. As Beagle 2 orbits Earth, it will have a relative

'Did anyone remember to pack Beagle 2 ?' Courtesy of Richard Slade.

velocity of around 17,500 mph (28,000 km/h). To escape Earth's gravity it must reach 25,600 mph (41,000 km/h). Travelling between Earth and Mars, the spacecraft has a velocity compared to the sun of nearly 90,000 mph because in addition to its own speed it has the velocity of Earth through space.

The final few days of the journey to Mars will be the most fraught. The Beagle 2 probe, now a spacecraft in its own right and not just a passenger, will be given a push of less than one foot per second on top of its velocity of Mach 31.5 (12,000 mph relative to Mars) and sent spinning, like a ball thrown by a rugby player, at around twelve revolutions per minute, into the atmosphere of Mars.

At an appropriate time after being slowed down, protected behind a heat shield, by the resistance of the atmosphere, to Mach 1.5 (1,000 mph), a mortar will deploy a pilot, or drogue, parachute whose function is to further brake Beagle 2 and stabilise the lander as it passes through the velocity of sound (Mach 1, roughly the same on Mars as it is on Earth). When the lander is falling at 230 mph, the drogue

COMPARATIVE VIEW OF
WESTERN HEMISPHERE
NORTH PACIFIC OCEAN
SOUTH PACIFIC OCEAN
NORTH ATLANTIC OCEAN
SOUTH ATLANTIC OCEAN
ARCTIC OCEAN
ANTARCTIC OCEAN
POLYNESIA
BRITISH AMERICA
UNITED STATES
SOUTH AMERICA
BRAZIL
Equator
Tropic of Cancer
Tropic of Capricorn
North Pole
South Pole
Level of the Sea
COMPARATIVE VIEW OF

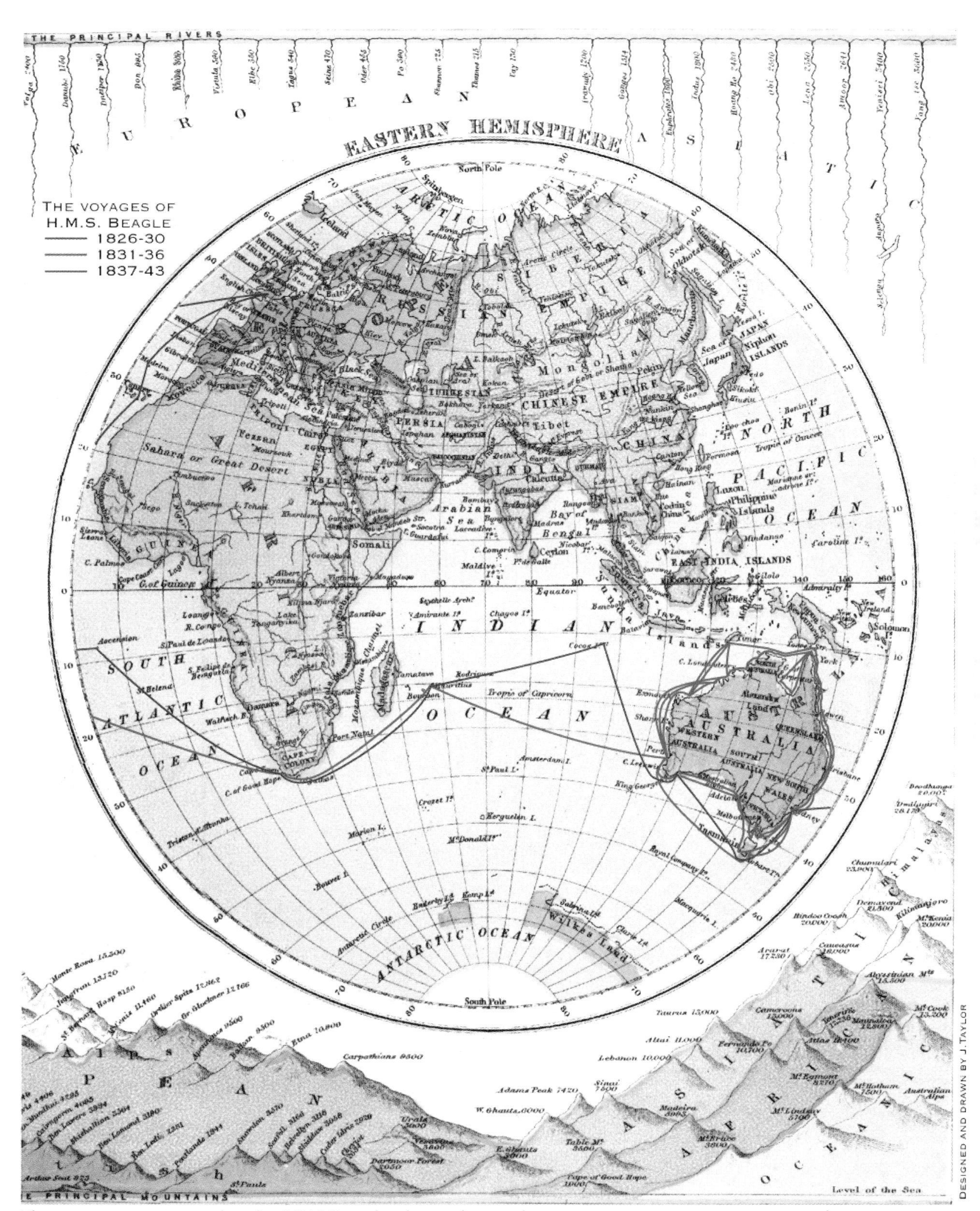

The various journeys taken by HMS Beagle *during her eighteen year career as a survey vessel.*

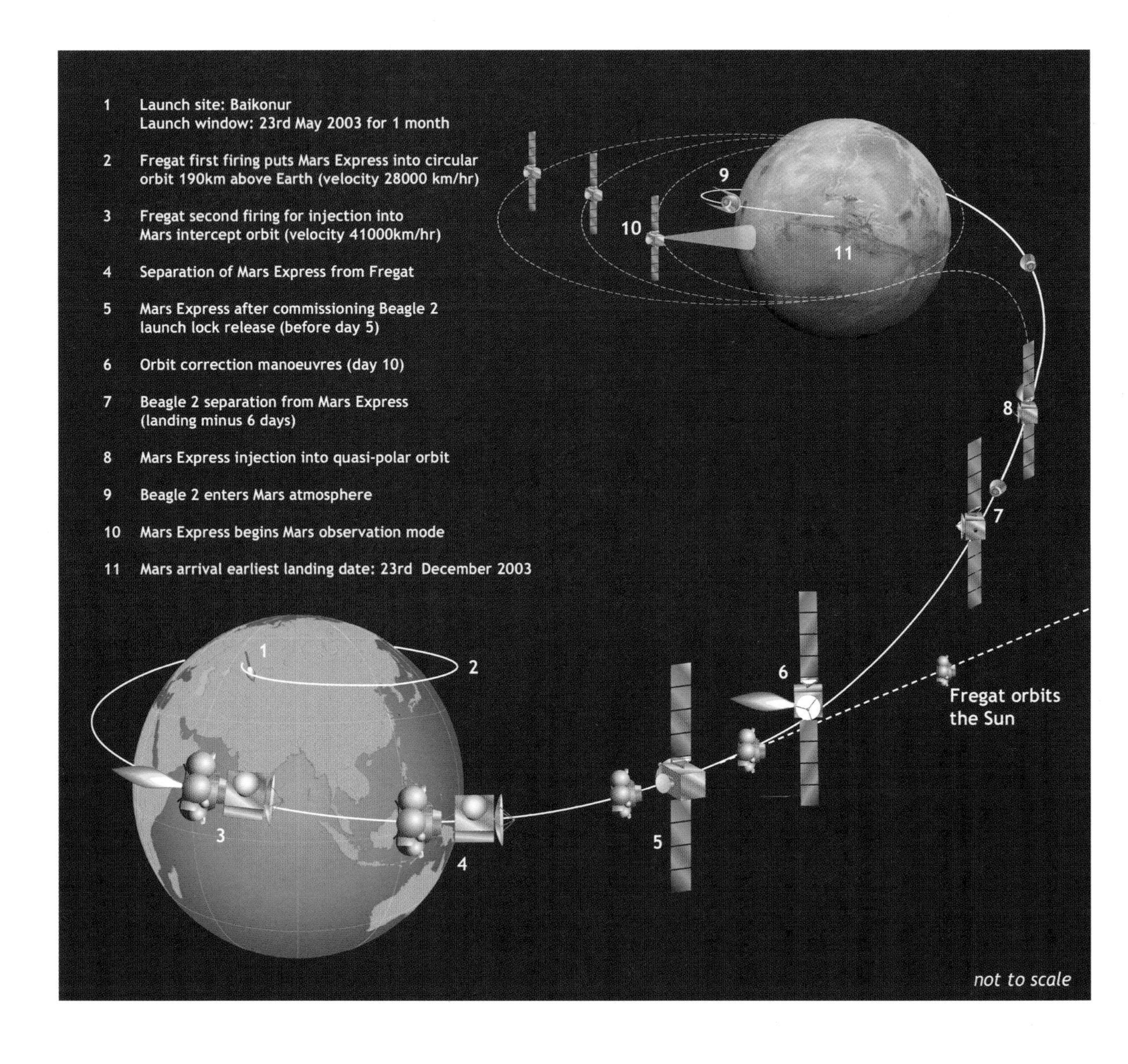

The various stages of Beagle 2's journey to Mars for launches any time after May 23rd 2003.

pulls the Beagle 2 probe into two halves, dropping the heat shield and removing the back cover. This action pulls the main parachute out of its bag.

The job of the main chute is to reduce the velocity to less than 40 mph. On contact with Mars, the lander, in its protective cocoon of inflatable bags, will bounce out from under the parachute until it comes to rest. It will drop the final metre on to the surface under the force of martian gravity. If it lands upside down the hinge which opens the lid is able to turn Beagle 2 over.

A journey of 250 million miles will be over, but unlike Darwin's *Beagle*, Beagle 2 will not be coming back unless a future generation of space explorers decides to collect it.

Time out

The perfection of a chronometer able to operate at sea over a period of nearly forty years enabled John Harrison finally in 1773 to win the Royal prize for devising a method able to establish longitude for navigational purposes. But even sixty years later, when HMS *Beagle* was sailing, the longitudes of many places around the globe were not accurately determined. The longitude of one of FitzRoy's first ports of call, Rio de Janeiro, for example was hotly disputed in 1832.

An Earnshaw chronometer similar to the one carried on HMS Beagle. *Courtesy The National Maritime Museum, London.*

To accomplish his tasks concerning longitude FitzRoy gathered on board twenty-two chronometers. Using this array of timepieces, HMS *Beagle* established, for the first time, a complete worldwide system of reference points for longitude measurement. FitzRoy listed his chronometers and rated their performance in the *Narrative of the Voyage of the Beagle*. He described how they were used in great detail: "The chronometers were embedded and permanently fixed for more than a month before *Beagle* sailed so sufficient time elapsed to test them".

They were "suspended in gimbals in a wooden box placed in sawdust, divided and retained by partitions, three inches thick". The sawdust came more than half way up the chronometer so that it could not move unless the ship was almost lying on its side: "Placed this way neither running on the deck, firing guns or running out of chains caused the slightest vibration". To confirm this fact FitzRoy put powder on the glass face of each chronometer and studied it with a magnifying glass.

Whilst even the Captain had to share his accommodation,

the chronometers had their own cabin; no one went in except to wind them daily at 9.00am or or to measure them. The eight-day models were wound on Sunday mornings.

The great majority of the twenty-two devices were not moved during the entire five year voyage. Only two of the chronometers were considered poor time keepers and their readings ignored.

To the second

Whilst it is travelling through space carrying Beagle 2, Mars Express will ensure that it is 'on an even keel' by monitoring the sun and a chosen star by tracker cameras. Any slight instability will be corrected by a jet of gas from thrusters.

But like a ship at sea, a spacecraft relies on a certain amount of dead reckoning to know where it is. Mars Express engineers will calculate how far it is on the way to Mars from knowledge of velocity and the time elapsed when a radio signal is sent and received. The velocity is established by measuring the Doppler shift of the radio signal sent back by MEx. This is the change in frequency caused when an object is moving away, like the sound of an express

The Beagle 2 probe attached to Mars Express showing the heat shield covered by the multi-layered insulation.

train that has just passed through a station. There are several opportunities during the journey of Mars Express where slight adjustments to velocity may be made to ensure that it is on the correct trajectory.

The radar altimeter trigger was tested by allowing a helicopeter fitted with the device to 'free fall' from fifteen hundred feet.

Before landing both MEx and Beagle 2, because the latter has no power system of its own, will be manoeuvred on to a crash course with Mars, aimed to enter the atmosphere at a shallow angle of sixteen degrees.

Several days out from the planet, at least five and maybe more, Beagle 2 will be ejected from Mars Express by a spring. Thereafter it will be on its own and dependent on a seven day timer.

The timer is helped by an accelerometer and the RAT (radar altimeter trigger) which together will tell Beagle 2 when to carry out various operations connected with the landing, such as when to fire the mortar to deploy the pilot parachute, or to fracture the bolts which hold the back cover and the heat shield together. Inflating the gas-bags is so critical an operation that the RAT has been included to detect when Beagle 2 is 200 metres above the ground. The gas generator, an explosive device, responds by filling the bags in around two seconds. The radar altimeter is a bit like the sailor in the crow's nest who sights land.

Once safely on the surface Beagle 2 reverts to relying on its timer. It will signal for the lace which holds the gas-bags together to be cut a discrete period after the accelerometer recognises that movement of the bouncing gas-bags has stopped. Other decisions, such as when to release the clamp band, which keeps Beagle 2 closed, so that the lid can be opened and the solar panels fold out, are all operations preprogrammed into Beagle 2 to happen automatically according to the clock.

With Beagle 2 nine minutes away from Earth even at the speed of light, and not in continuous communication, it simply is not possible to command, in real time, each and every activity, with a signal from the team on Earth. Lots of other procedures during the scientific analysis, particularly using the gas analysis package, will depend on computer programmes involving Beagle 2's timer chip.

Starting and stopping

The power to drive a ship of course comes from her sails. Very full lists of the complement of sails of HMS *Beagle* exist because both FitzRoy and his successor placed orders for

Sewing the sails for The Ship that Sailed to Mars by William Timlin.

new sets before leaving England. The language of the sail maker is completely opaque to the uninitiated; it is a world of royals, trysails, jibs, gallants, spankers, staysails and courses.

The modern-day equivalent for Beagle 2 is the parachute which provides the stopping power for the spacecraft. The jargon is every bit as mystifying. Beagle 2's pilot chute is a type known as a disc-band-gap and the main parachute is a ring-sail. A ring-sail parachute is almost exactly what is says, a ring of small, almost sail-like, panels which billow out to spill the air which is passing through the parachute during the slowing down process.

Traditionally parachutes are increased in size in multiples of four so Beagle 2's main stopping power is provided by eight rings of twenty-eight panels making two hundred and twenty-four in all. The load of the lander is shared through the parachute by twenty-eight rigging lines, which continue into the seams holding the parachute fabric panels together. The lines are the thickness of thread but have enormous breaking strain. The final single strop is about as thick as parcel string with the difference that it is able to support 1800 kilograms and could be used to tow a car.

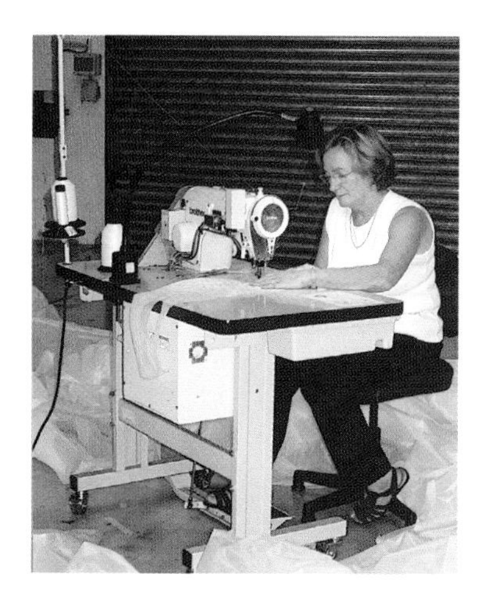

Work begins on the Beagle 2 parachute at Lindstrand Balloons factory.

Interestingly the fabric used in the parachutes comes from the same sources that supply sails for yachts today, synthetic materials rather than cotton, but still a good link to the days of HMS *Beagle*.

FitzRoy attached the utmost importance to HMS *Beagle*'s rigging. At his direction the rigging ropes were larger than usual, chains were sometimes used to give greater strength. "In no place was there a pulley or block that was not matched to its rope or chain". The use of chains to secure spars to the masts was another of FitzRoy's innovations – it did not become common practice in the Royal Navy until at least ten years after the HMS *Beagle* circumnavigated the World with Charles Darwin.

Good vibrations

A sailing ship in bad weather is not the best place to try to operate delicate instruments. It should not then be surprising that it took years to develop chronometers capable of keeping time accurately enough to measure a ship's longitude. Travelling aboard a spacecraft, equipment and

Parachute test, dropped from a hot air balloon over Slepe Airfield, Shropshire.

instruments experience an even rougher ride. Everything aboard Beagle 2 has to be 'qualified' to survive the regime it will encounter during the mission.

The most stressful time will be during launch when the vibrations from the giant rocket engines threaten to shake Beagle 2 to pieces. So each part of Beagle 2 has to be tested by vibrating it in all three dimensions and randomly at levels over and above what might be experienced on take-off. Delicate accelerometers, the equivalent of FitzRoy's powder, monitor the effects at various points. Some elements of Beagle 2 such as the robotic arm and the spin-up and eject mechanism, have to be free to move during the mission; for launch they will be held down with bolts which will subsequently be broken by passing an electric current through them; expansion fractures the bolt at a notch.

There are a number of other nasty experiences coming to Beagle 2 including the noise of the take-off, the egress of the air as the rocket goes from atmospheric pressure to the hard vacuum of space and the shock of landing in the gas-bags and falling out of them on to the surface of Mars. Imagine pushing your computer off the desk on to the floor and expecting it to be unaffected. All the above conditions have been part of Beagle 2's qualification programme.

Send us a postcard

FitzRoy expected Darwin would quit HMS *Beagle* at the first landfall. But the opportunity did not arise for some weeks. Passing through the area which used to be known to shipping as Finisterre but was recently renamed FitzRoy in honour of *Beagle*'s captain, the ship sailed across the Bay of Biscay and on past its intended destinations of Madeira and the Canaries to St Jago in the Cape Verde Islands.

Picture postcard scene of Cocquimbo painted by Robert FitzRoy. Courtesy UK Hydrographic Office, Taunton. Crown Copyright.

Beagle *would have welcomed the modern postal service – a first day cover commemorating the hundred and fiftieth anniversary of Darwin's visit to Australia.*

As they left this haven on February 8th, 1832 Darwin wrote to his father saying he was hopeful of "meeting with a homeward bound vessel somewhere about the Equator – the date will tell whenever the opportunity occurs". Two days later they met a ship of some sort but it was not heading to England; Darwin thought the conveyance insecure and did not commit his letter to it. Instead he waited until arriving in San Salvador, Brazil on February 28th by which time he had added more information making the letter an epistle for the whole family. They finally received the communication in Shrewsbury on May 3rd.

Darwin only got word that it had arrived when he found a response from his sister Susan awaiting him when he arrived back in Montevideo from Bahia Blanca on October 24th. No wonder Captain FitzRoy had to be the law and God, there was no way he could write home for advice. If he needed a superior's order he had to find a higher ranking officer, perhaps on a British ship he met, or stay put whilst the message went back and forth to England.

Will Beagle 2 please call home

You could travel to Mars in the time it took Darwin to find out that his first letter had reached home. Beagle 2's telecommunications will be a bit faster but still incredibly slow in the modern era of telephone, fax, e-mail and text messaging. A round-trip radio signal from Mars to Earth will take about fifteen minutes when Beagle 2 lands. As Earth continues round the solar system and leaves Mars behind the delay will get longer and longer. By October 2004 it will not be possible to send messages at all because the two planets will be on opposite sides of the solar system and the Sun will be in the way.

Nor will it be possible for Beagle 2 to send messages to Earth and receive them back just when it likes. To start with Mars turns on its axis with a period of twenty-four hours and thirty-seven minutes so the side of Mars where Beagle 2 lands will face away from Earth at least half the time. Secondly, Beagle 2 does not have a powerful enough transmitter to communicate data direct to Earth, it uses the Mars Express orbiter as a relay. Mars Express will be orbiting Mars with a period of just over eight hours, as long as it is not manoeuvring it sees Beagle 2 landing site two or three times a day.

One of the most inconvenient times for Beagle 2 is just after it has landed on the martian surface. There may not even be time to send an "arrived safe" message to the orbiter which will be flying behind the lander and passing over the landing site less than a minute after the landing event. Beagle 2 takes several minutes to open its lid so although the orbiter can collect signals from a thirty degree angle away from the

NASA's Mars Odyssey spacecraft, one of the routes by which Beagle 2 can signal back to the waiting scientists.

vertical it is likely to be out of range. If Beagle 2 fails to make this contact with its companion then it does not see MEx for at least another ten days as the orbiter is busy getting itself into the desired orbit.

The Lovell Telescope at Joddrell Bank Observatory will be able to detect whether Beagle 2 has landed but is unable to send messages.

Another way of communicating with Beagle 2 is to send a message via a passing spacecraft. Fortunately there is one, the NASA Mars Odyssey which is already in orbit around Mars. Odyssey will be fully committed to NASA's twin Mars exploration rovers from early January 2004 onwards but a deal has been struck which provides for mutual cooperation in communications between the American and European orbiters and their landers.

There is just one more way of finding out whether Beagle 2 is alive and functioning, and that is using Jodrell Bank. The giant telescope is undergoing modifications to enable it to listen out for Beagle 2 immediately after landing. Fortunately there will be a few hours a day in the early stages of the mission when Britain's best known telescope and its equally famous tiny spacecraft will be staring at each other across the wastes of space.

It is all a bit reminiscent of Darwin trying to find a homeward bound ship in the middle of the Atlantic Ocean.

St Jago's Rocks

Perhaps one of the reasons why Darwin did not leave HMS *Beagle* at the earliest possible minute was because he had a new passion – geology. FitzRoy wrote to Francis Beaufort "a child with a new toy could not have been more delighted than he [Darwin] was with St Jago". Looking at a band of sea shells incorporated into lava high above sea-level, Darwin appreciated that a sequence of events was being recorded "It then first dawned on me" he wrote "that I might write a book on the geology of the various countries visited".

"There is nothing like geology; the pleasure of the first days partridge shooting or the first days hunting cannot be compared to finding a fine group of fossil bones."
Charles Darwin, in a letter to his

Darwin had previously had a little field geological experience in North Wales with Adam Sedgwick, Professor at Cambridge, and he started to read Lyell's *Principles of Geology*, recommended by Henslow (who added "on no account to accept the views therein advocated") and bought for him as a leaving present by FitzRoy. It was advice that went largely unheeded since the message that Charles took from the book was the importance of the geological timescale.

Because Darwin wrote *On the Origin of Species*, a zoological work, his, and thus HMS *Beagle*'s, contributions to geology are often forgotten. The first paper Darwin prepared on his return was about rocks and read at the Geological Society, who elected him a Fellow long before the zoologists got round to it. Some of his most sought-after specimens were the giant fossils he recovered whilst trekking from one side of South America to the other.

When he returned home he arranged for them to be described by Richard Owen, Professor at the Museum of the Royal College of Surgeons. They turned out to be "treasures", Darwin's word. Surprisingly Owen later became an opponent of Darwin's ideas on natural selection.

Land ahoy

Beagle 2's first and indeed only port of call on Mars will be Isidis Planitia within an ellipse centred on 11.6 degrees North and 90.5 degrees East.

When Mars is at opposition in 2003, the closest for thousands of years, it will be possible to see Isidis Planitia relatively easily with a 150 times magnification telescope. It will be in the lower left quadrant (astronomers look at things upside down) to the left of the Syrtis Major, the dark feature most easily identifiable on Mars.

Isidis Planitia

Isidis is a thousand kilometre wide crater believed to be infilled with sedimentary debris washed down from the Southern Highlands. It is just north of the equator and meets the engineering constraints for a landing site. Its elevation is below the martian datum (an artificial sea-level defined by atmospheric pressure) and thus is appropriate to provide the maximum stopping power for the parachute. The location is not too far north so the angle of the Sun is high enough to allow good conditions for battery charging from the solar cells. The abundance of rocks is about fourteen percent ground cover. Not enough to endanger the landing systems and frighten the engineers but sufficient to keep the geologists happy.

Isidis Planitia is a basin with sedimentary infill dating from the late Hesperian and Amazonian times. There is no absolute geologic timescale for Mars, only a stratigraphic one and a very basic one at that. There are three main epochs in time order: Noachian, Amazonian and Hesperian. Relative ages

Topography of Mars from the Mars Orbiter Laser Altimeter (MOLA) on NASA's Mars Global Surveyor orbiter. The highlands are shown in yellow to red, low altitude areas in blue. The pale blue area is Isidis Planitia, the landing site for Beagle 2. The target ellipse for the landing is around three hundred kilometres long by eighty wide, an area about the size of England south of the M4 motorway.

are worked out on the basis of the number of impact craters above a certain size in a given area, a million square kilometres. Some authors have tried to guess the absolute ages of locations on Mars by comparison to the Moon. Isidis has a crater count of about 0.75 of the average lunar Mare which suggests an age of around 2.7 billion years.

The NASA Mars Global Surveyor has provided detailed images of Isidis showing the presence of volcanic cones which are typically between 400 and 900 metres in diameter. These were probably produced just below or even at the surface by the interaction between magma and volatiles such as water or carbon dioxide. The cones have been degraded to differing extents indicating that some are much older than others. Whilst many of the youngest cones are in the west of the region, ones of different ages do occur in close proximity. The evidence of persistent and repeated volcanic activity and the associated volatiles indicates that Isidis Planitia is indeed an exciting place for Beagle 2 to be carrying out its scientific investigations.

Conversations between planetary scientists and astronomers always get confused because of image inversion. Sailors also used to make mistakes muddling the similarly sounding starboard and larboard for the opposite sides of their ship. FitzRoy insisted on the use of the now commonly recognised port for the left side of the ship as viewed facing forward.

It is requested that the word "Port" may always be used instead of "Larboard"
from the FitzRoy Book of Orders by John Lort Stokes copied from FitzRoy original c. 1831, property of the National Maritime Museum, London.

Orders are orders

FitzRoy received his orders from the Admiralty Office in a long letter dated November 11th 1831. It began by informing any senior officer who may fall in with Commander FitzRoy against deflecting him from his important duties. It went on they should not "in any way interfere with him or take from him, on any account, any of his instruments or chronometers".

'The Orders' gave the Captain of the *Beagle* a long list of places, both in the Atlantic and Pacific, where it wanted checks for longitude using the chronometers. To rate the chronometers, astronomical observations of Jupiter's third and fourth satellites should be made, noting when they went behind the planet and emerged again.

Surveying tasks included measurement of depth of water over submerged hazards and information about potential harbours: "it is of immense consequence to a vessel which has lost her masts, or a large part of her crew, to have a precise knowledge of a port to which she is obliged to fly". In this respect the west coast of South America was least well known. FitzRoy was not to go into excess detail however: "there will be no time to waste on elaborate drawings. Plain distinct roughs, everywhere accompanied by explanatory notes in the margin will be of far greater value ... all charts and plans should be accompanied by views of the land". A specific task related to trying to work out how coral reefs were formed. In due course Darwin would write a paper on this subject.

Information was required about tides and the weather: no day should pass without a series of measurements of compass bearings, magnetic dip, barometer readings "to the third place of decimal" and "the extremes of the self-registering thermometer should also be recorded". Use of Beaufort's new wind scale was recommended.

It was generally accepted that naval vessels would make collections of flora and fauna, and record the resources other ships might expect at various places "the chief productions that can be obtained [from the natives] and the objects most anxiously desired in return".

As if all this was not enough, the officers should document any remarkable astronomical phenomena: "eclipse, occulting and moon culminating stars will furnish those chances in abundance" and "if a comet should be discovered its position should be determined every night by observing".

Information on all subjects touched on in this memorandum should be transmitted back from time to time to their Lordships at the Admiralty "so that if any disaster should happen to *Beagle*, the fruits of the expedition may not be lost altogether".

If a specific set of orders were to be issued to space missions visiting Mars then they would almost certainly contain instructions about finding water. NASA's mantra in respect of exploring the planet is "follow the water", they suggest this approach may indicate evidence of life on Mars.

Cartoon depicting how thinking has changed concerning the possibility of water on Mars. Thanks to Teemu Mäkinen, courtesy Finnish Meteorological Institute and ESA.

The laboratory on Mars

Beagle 2, like its predecessor, HMS *Beagle*, will address a wide range of subjects – the life question, of course, but also geochemical and mineralogical information for the rocks and the chemistry of the martian atmosphere.

The full list of objectives

A geomorphological characterisation of the landing site.

A geological investigation of the nature of the rocks found there, especially their light element chemistry, composition, mineralogy, petrology and age.

An investigation of the oxidation state of the martian surface, both within rocks, soil and at protected locations beneath boulders.

The presence or recognition of any or all of the following – water; appropriate inorganic minerals showing that water may have been present, e.g. carbonate; carbonaceous debris; organic matter of complex structure and lastly, isotopic fractionation, between organics and carbonate, that could be an important indicator of life.

A full composition of the atmosphere for the purposes of establishing the history of the planet and processes involved in seasonal climatic changes or diurnal cycling.

A search for trace atmospheric gases which would be indicative of extant life at some location on Mars.

This list of objectives in many respects exceeds what has been attempted or achieved by previous missions. It goes far beyond the goals stated for the other landers travelling to Mars in 2003.

None of the objectives provides an unambiguous answer alone about life on Mars. It is therefore the key objective of Beagle 2 to perform the whole suite of experiments using an integrated package, at the centre of which is a mass spectrometer capable of several modes of operation to analyse gases from various samples, collectively known as the Gas Analysis Package, GAP.

Cameras mounted on the lander will be used to select the most appropriate materials to be investigated such as soil from depth below a boulder and specimens from the interior compared to the exterior of rocks, first classified by chemical, mineralogical and microscopic techniques. The results will be considered together to shed light on what, in the light of recent findings on the martian meteorites, remains a crucial open question of fundamental importance: were there ever conditions on Mars appropriate to life and does any trace signature remain?

With the exception of the gas analysis package, Beagle 2's instruments are carried on the PAW. From left to right: wide angled mirror, camera, corer/grinder, Mössbauer spectrometer, microscope, camera and X-ray spectrometer.

Like HMS *Beagle* it is important not to lose the information. Beagle 2 will communicate with Earth every day, both to send results and to receive new orders.

Flash, bang, wallop what a picture

Although H.G. Wells was responsible for one of the most successful martian scare stories, he also wrote light romance. One of these, drawing on Wells's early experiences, is *Kipps*, the life and loves of a poor draper's assistant. It was turned into the musical *Half a Sixpence*, a song from which suggests that "there must have been a photographer to record the happy scene" at all the events in history. Of course there was not, the photographic process was only invented in the 1850s, after the completion of all of the third *Beagle*'s voyages.

But HMS *Beagle* did need to record events and Captain FitzRoy took along an artist or draughtsman, as his job was officially called, for the purpose. First in the post was Augustus Earle who joined the ship at the same time as Darwin in Devonport. He had previously journeyed to New Zealand and wrote a book about his experiences which was published during the first year of *Beagle*'s trip.

When Earle's poor health forced him to leave the *Beagle* voyage in August 1833, he was replaced at Montevideo by Conrad Martens. The work of Martens aboard the *Beagle* is much better known. Darwin in a letter home once had to confess to his father that he had "been a little extravagant and bought two watercolour sketches from Martens for three guineas each".

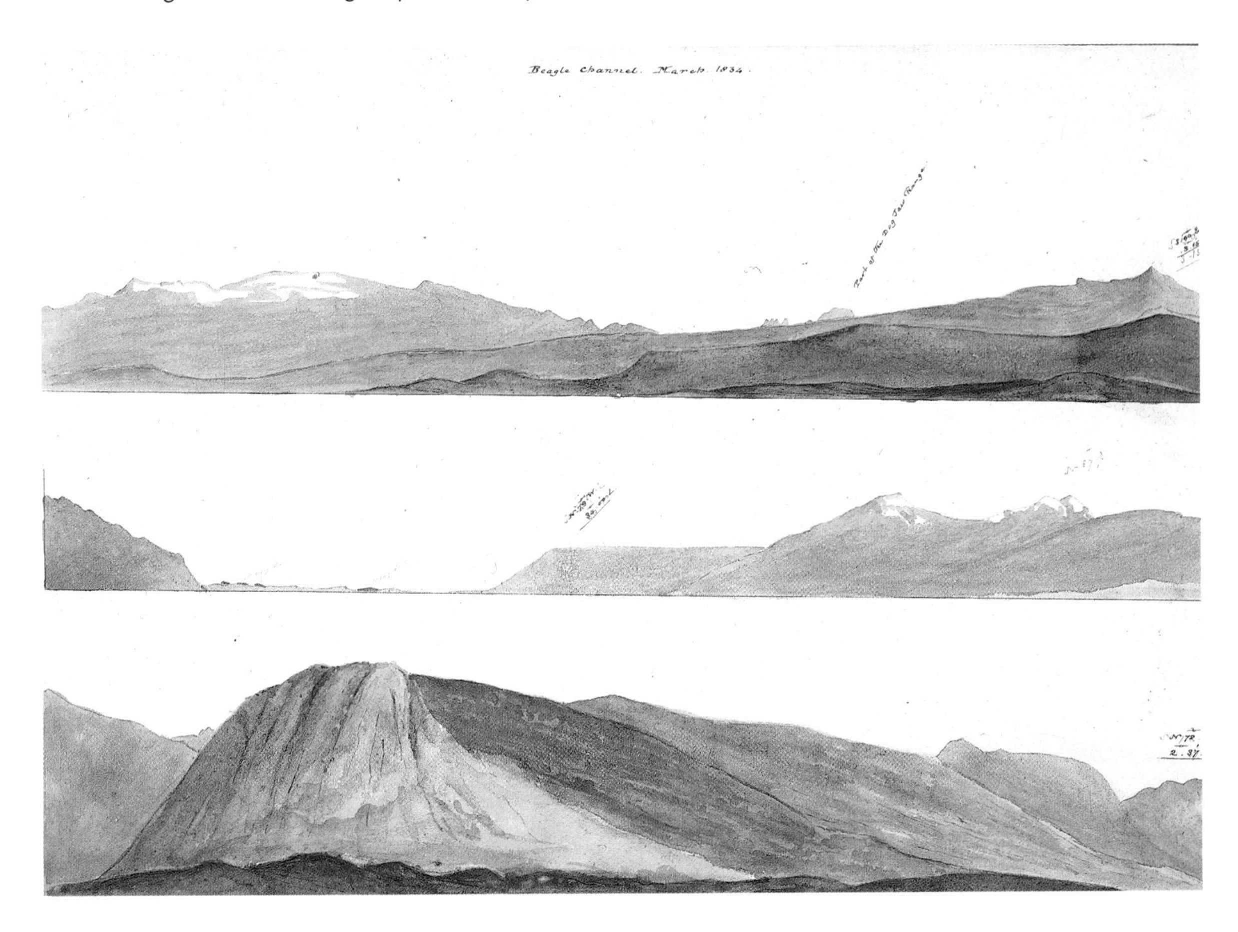

Views of the Beagle Channel by Conrad Martens. Such pictures were appended to HMS Beagle's charts. Courtesy UK Hydrographic Office, Taunton. Crown Copyright.

Although individual pictures by both Earle and Martens are often referred to, one of the main tasks in the draughtsman's day was to paint the coastline. The continuous views of the shore were a vital part of the charts produced as the voyage unfolded.

It seems after FitzRoy was forced to sell *Adventure*, *Beagle*'s companion vessel, there was less room for the artist who subsequently spent more time on land.

During the third voyage of the *Beagle*, there does not appear to have been an official draughtsman but John Lort Stokes, Bartholomew Sulivan, Philip Gidley-King and L.R. FitzMaurice were all passable artists.

A picture is worth a thousand words

A necessity for any space mission like Beagle 2 is a camera, after all a picture is worth a thousand words. In Beagle 2's case it cannot do anything without its cameras, they are the eyes that will govern the activities on the surface.

During the landing all the operations of the spacecraft will be pre-programmed in, including opening the lid and folding out the solar panels. The opening operation will allow a small convex mirror to pop up, this should enable one of the cameras, which will still be in a stowed position, to get an initial 'picture postcard' view of the martian surface to signal home during an early, hopefully the first, communications session.

The author, Dave Rowntree and co-workers pictured by one of Beagle 2's cameras during development. With thanks to Jean-Luc Josset.

As soon as it is able, Beagle 2 will raise its robotic arm to its full height, ninety centimetres above the lander base, and rotate it through 360 degrees taking a series of pictures which can be stitched into a complete panorama of the landing site. So Beagle 2's cameras will be used, not just for snapshots of the landing site, but to survey it, just as HMS *Beagle*'s artists carefully mapped the coast of the places the ship visited. Bearing in mind the urgent need to get back information, images can be compressed by computer software routines so that the communications channels are not blocked by the huge quantities of data needed to provide the highest resolution images.

Because there are two cameras, nineteen and a half centimetres apart, slightly toed in to overlap, it will be possible to have a three-dimensional or stereo image. The digital data will be processed by computer software and fed into a system which will create a solid model of the landing site by a process called rapid prototyping. The model will be used to test the operations of the various tools which Beagle 2 will use to recover samples and the cameras will record progress.

The view seen in the wide angle mirror just prior to closing the lander lid. The next picture to be taken will be on Mars.

A second purpose of the convex mirror is that it can be used to see what is going on when the mole is being deployed from its carrier or retracted after sampling. Beagle 2 can look behind itself without wasting time and power moving the arm.

As the mission progresses the cameras will perform their own scientific functions, photographing at different wavelengths using twelve different coloured filters.

It is also planned to use the camera for a number of astronomical observations, and to investigate the dust burden in the martian atmosphere.

Mole

To be able to address some of its most important scientific objectives, Beagle 2 needs to grab some Mars soil. Several such samples will be analysed for organic matter, mineral

An early photo-opportunity for the mole in a quarry at Sandy, Bedfordshire, thanks to Lafarge Aggregates.

components and isotopic ratios to seek evidence of extinct life on the planet. The means to acquire soil samples is the mole, PLUTO, which is deployed somewhat like a dog on a leash for several sorties away from the lander. The mole can travel laterally across the surface as well as directly downwards into the soil.

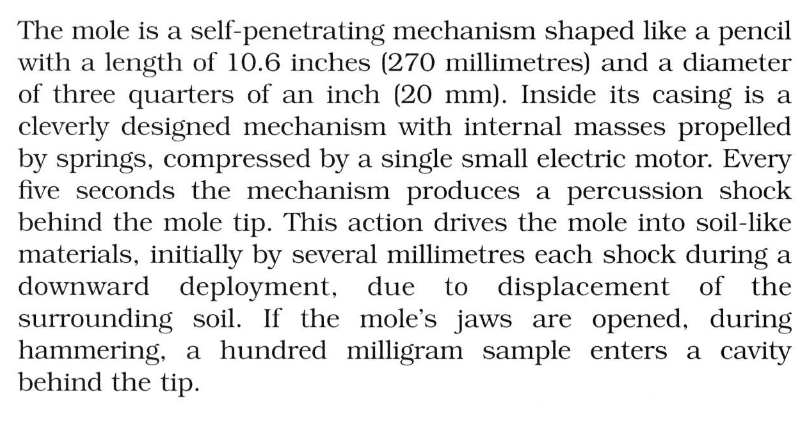

The mole is a self-penetrating mechanism shaped like a pencil with a length of 10.6 inches (270 millimetres) and a diameter of three quarters of an inch (20 mm). Inside its casing is a cleverly designed mechanism with internal masses propelled by springs, compressed by a single small electric motor. Every five seconds the mechanism produces a percussion shock behind the mole tip. This action drives the mole into soil-like materials, initially by several millimetres each shock during a downward deployment, due to displacement of the surrounding soil. If the mole's jaws are opened, during hammering, a hundred milligram sample enters a cavity behind the tip.

During tests in the laboratories at DLR Cologne, the mole reaches the bottom of the container that Lutz Richter is standing by.

Because it moves away from the lander, the mole needs to be connected to the PAW by a tether which is paid out as PLUTO goes about its business. The tether doubles as a means of providing power to the internal mechanism, transmitting command signals and sending back data. The mole can move laterally across a smooth surface uninhibited, however if it encounters a large rock, it is unable to go over or round and instead is deflected downwards. This means that it can burrow well below the surface and collect a sample from somewhere that is protected from the harsh surface conditions where organic residues may have survived. The permanently shaded location under a boulder could have trapped out water. In either case, moving across the surface or burrowing down below it, PLUTO returns to its home base by way of a winch on the PAW reeling in the connecting tether aided by shocks in the reverse direction.

The surface

From what is known from previous missions about the martian surface, it is likely that the topmost several metres in most regions of the planet are indeed 'soil' and not solid bedrock, allowing PLUTO to be a viable means of providing access to subsurface material. The Viking landers of twenty five years ago only scratched the surface and turned over a small rock in their quest for organic molecules. Viking scientists were careful not to say that there were no organic chemicals present, only that they did not find them and offered the environmental conditions as a possible reason why such molecules might have been destroyed.

The mole has nautical similarities when it comes to recording how far it has travelled. The cable drum which pays out the mole's tether keeps a record of the number of revolutions as a measure of distance just like the knotted line. The same method will record the depth to which the mole has burrowed, just like the ship's depth measure, the line thrown overboard with a plumb weight attached to settle to the bottom.

The mole speed, though, is a bit too slow to measure in knots but its penetration limit below the surface is about a fathom (six feet or nearly two metres).

Don't let them grind you down

It is imperative for the scientific aspirations of Beagle 2 that the weathered outer surfaces of rocks are removed. NASA's Pathfinder mission in 1997 had no provision for eliminating the red dust which covered all the boulders, so frustrating chemical analysis and detailed photography. Some ingenious data processing tried to strip off the soil layer but still all the rocks gave similar results even if they were of different types.

To combat the problem of dust and an obscuring outer rind, Beagle 2 is equipped with a corer-grinder. It was designed and built by a dentist. The same device performs all the

Samples from the mole or corer/grinder are delivered to the Gas Analysis Package (GAP) by dropping them into a funnel opening on the deck of the lander.

functions. To grind off the outside layer, the drill bit is allowed to wander across the surface taking off a few millimetres below which the rock should be less weathered. Once the device is locked, the drill bit, which is hollow, bites into the rock and cuts an annulus or ring with chips of the sample entering into the central cavity of the drill bit. After penetrating to a depth of about one centimetre, the drill bit is reversed and any solid length of specimen is snapped off and all fragments are grabbed. About sixty milligrams (a couple of pinches of salt) can be picked up for delivery to the gas analysis package for study.

The blow pipe

Darwin equipped himself, for his geological effort, with a hammer, the equivalent of the corer-grinder, and a blow pipe, a tool for studying the elemental makeup of samples. He mentioned in his diary putting the latter to use on those first seashells he saw thirty or forty feet above the shore on St Jago.

A blow pipe sounds like a weapon that primitive warriors might use to fire darts at marauding explorers such as *Beagle*'s naturalist. Not so, it is an early means of conducting chemical analysis. Darwin appears to have been interested in its application from boyhood since a school friend recollected him using one at Shrewsbury.

The idea behind the blow pipe is incredibly simple, the hard part is mastering its use and interpreting the results. It consists of a tube attached to an angled metal pipe with a narrow orifice at the tip. The tube is placed in the user's mouth and the tip inserted in a sooty flame (Darwin would probably have used a candle on board HMS *Beagle*). When the analyst blows steadily into the flame, a much hotter and direct jet, rather like that generated by a Bunsen burner or a gas torch, is produced. By holding tiny fragments of rock in different parts of the flame it is possible to perform a number of tests. The difficulty was keeping up a steady supply of oxygen from the lungs for long enough; Darwin would have had to master circular breathing, that is keeping a reservoir of air in his mouth whilst refilling his lungs through his nose.

The easiest test using a blow pipe was to see how readily bits of rock or crystals could be melted. Darwin would have selected thin, lath-like fragments and studied the rounding of their edges in the flame. Whilst some minerals such as quartz are impossible to melt with a blow pipe, others, held in forceps, rapidly collapse into a bead.

Other tests Darwin could easily have done would be to sprinkle crushed rock powder through his blow pipe flame. He would then have looked for the characteristic colours which many elements generate, for example intense yellow from sodium, green from copper or crimson from strontium. A blow pipe expert would have had blocks of charcoal and plaster of Paris and using these accessories could react various minerals to create characteristically thin films. For example, heating a gold ore on a block gives a metallic film of gold.

A blowpipe, with its trumpet-like mouthpiece, courtesy of the Whipple Museum of the History of Science, Cambridge.

The blow pipe flame has the very advantageous property of being both oxidising (capable of adding oxygen) or reducing (taking it away). The oxidising region is at the most extreme tip. Placed here a sample of pyrite, iron sulphide would have its sulphur converted to sulphur dioxide and Darwin would have been able to detect its characteristic pungent odour.

Alternatively samples of oxidised minerals, for instance rusty iron, could be reduced in the region of the flame just in front of the hottest part. This position contains unburnt carbon monoxide, produced from organic carbon in the candle wax. Carbon monoxide rapidly reacts with the oxygen from iron compounds such as rust to give black ferrous oxide.

It is not possible to detect carbon in rocks with a blow pipe. But some of the other elements and minerals which Beagle 2 will study on Mars could be investigated with the device which journeyed with HMS *Beagle*. The martian soils may contain sulphur, which would be readily detectable in oxidising conditions, whereas with a reducing flame it would be possible to show that the red colour of Mars is due to ferric iron compounds.

X-raying Mars

Although the primary task of Beagle 2 is to search for evidence of past life on Mars, it will also carry out a full geological characterisation of the landing site to make sure it can consider any results it gets in context. To study the chemistry of the rocks on the red planet, Beagle 2 carries an X-ray fluorescence spectrometer. This will acquire information on major, minor and trace elements by identifying lines of characteristic energy in the spectrum of X-rays emitted when the samples are excited by radioactive ^{55}Fe and ^{109}Cd sources.

A number of elements having a wide spread of X-ray energies are included in the calibration target; for example, aluminium, copper, molybdenum, titanium and manganese. These give a scale for identifying the elements present on Mars. The abundance of the elements is obtained from the strength of the X-ray signal; the calibration target provides a reference point because quantitative data have been acquired on Earth.

The data from the X-ray spectrometer as well as from some of the other instruments will be gathered from soils and rocks which have had the weathered surfaces removed by the grinder. The PAW on the moveable arm is able to position an instrument like the X-ray spectrometer precisely to the point where the weathered surface has been removed.

Why is Mars red?

To study the mineralogy of martian rocks Beagle 2 carries another kind of spectrometer which measures how gamma rays interact with iron nuclei in the various minerals. The property, called the Mössbauer effect, is extremely useful in gaining information about the make-up of rocks because almost all the common rock-forming minerals contain iron, but in a variety of states.

The Mössbauer spectrometer will reveal the answer to an age-old question "why is the red planet red?". It must be because iron compounds are changing their oxidation state (rusting) but the exact nature of the process is unknown and of course it is not clear yet whether martian rocks are oxidised all the way through. The Mössbauer instrument needs, before it starts its work on Mars, to be checked against standard 'spots' of the Damien Hirst calibration target, which contain many different forms of iron. Interpreting Mössbauer data requires calculating how the signal is made up from many possible mixtures of components. Understanding how the instrument is performing in respect of known standards makes interpretation possible.

The Mössbauer spectrometer will be used to compare weathered surfaces and those for which the weathering has been removed by the grinder.

Looking inside rocks will also be a first for martian exploration. If the interiors are unoxidised then the chemical evidence of past life on Mars may have had an opportunity to survive.

A rusty planet.
From The Ship that Sailed to Mars by William Timlin.

Goniometer

Apart from using chemical analysis to identify minerals, Darwin was equipped with goniometers. With these he was able to measure the angles between crystal faces which are characteristic for individual minerals.

In the summer of 1832 Darwin wrote to Henslow that he was making notes on his rocks and followed up with another

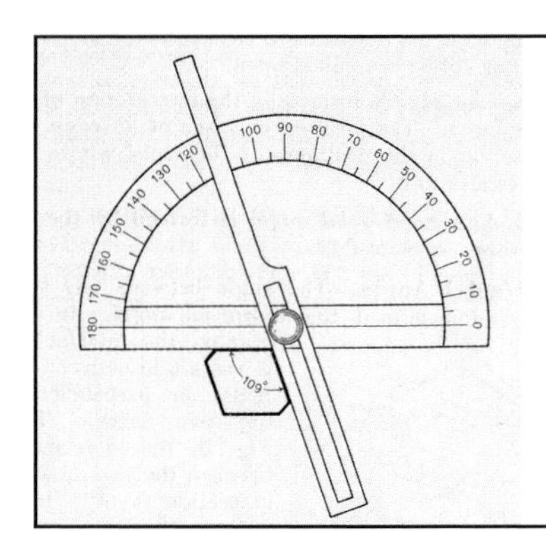

Darwin's instruments

The simplest kind of goniometer was a type of protractor in which the angle between crystal faces could be measured by adjusting a moveable arm and reading off a scale. A more exact device, called a reflecting goniometer, relied on the reflection of a distant source of light off adjacent crystal faces and measurement of the movement of the viewing optics. Typically the source of light was daylight from a window and the image of the glazing bars was looked for.

letter in November the next year asking Henslow for advice and a book on how to operate his reflecting goniometer. Henslow replied that use of a goniometer was advised only for "an experienced mineralogist in his closet".

Despite a discouraging response Darwin went ahead with his plan. Alfred Harker, distinguished Professor of Mineralogy at Cambridge, whose department inherited many of Darwin's rock collections, said in 1907 that the naturalist's notes were "a monument of patient labour". Each description contains information as seen by the eye, a hand lens, beside notes on locality and occurrence. "On the opposite page are additional notes, also made during the voyage, giving the results of examination made with the blow pipe, goniometer, magnet and acid bottle." The latter would have been Darwin's way of recognising carbonate minerals. The technique was employed on the Nakhla meteorite but unfortunately cannot be used on Beagle 2.

The example which Darwin set with his collection is being followed carefully by the Beagle 2 team. Unlike Darwin, the team will not have a box of samples to study at leisure in the laboratory. It is still vital however to characterise fully, by different methods, every rock or sample which is being studied for evidence of ancient life. It will then be possible to put any results obtained into context, even if the laboratory study is taking place many millions of miles away on another planet.

Seeing is believing

Charles Darwin swore by his microscope for examining the tinier creatures he found on his travels. He rated working with it second only to geology. It was his way of identifying "new and curious genera".

One of the microscopes taken by Charles Darwin on HMS Beagle, *Courtesy English Heritage.*

Charles received the first microscope he owned as an anonymous gift (later found to be from a Cambridge contemporary, John Maurice Herbert). Herbert described his present as a Coddington; John Henry Coddington was a tutor at Trinity College Cambridge who made improvements to a microscope built by the London Instrument makers, George and John Cary. The Coddington was packed in a box approximately six inches by three inches.

Another microscope that Darwin had on board HMS *Beagle* was supplied by Bancks and Son, 119 New Bond Street. This device was recommended to Darwin by Robert Brown FRS, who was librarian to Sir Joseph Banks, the President of the Royal Society who initiated the meteorite investigation. Brown was famous for discovering Brownian motion (the propensity of tiny particles to move in a drop of fluid).

After being a year or so on HMS *Beagle*, Darwin realised the importance of having his specimens well lit in order to see the detail. He wrote home from Montevideo, asking that he be found a large lens on a stand "with plenty of motions" so that he could "have steady light on an opate (sic) object". He suggested that his brother Erasmus be charged with the task of finding an optician who "must have made some such contrivance". Whether the accessory ever arrived is not known.

Under the microscope

Beagle 2's microscope turns out to be not much smaller than either of Darwin's. Its overall length is 4.4 inches by 1.8 inches wide at its maximum diameter.

Testing the Beagle 2 microscope on Earth:

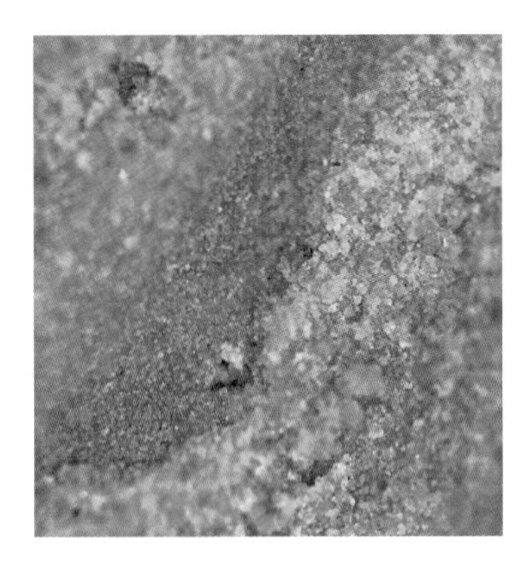

Layered limestone (cave sediment) from Piz Alv, Switzerland.

Basaltic volcanic rock from the Cady Mountains, Mojave, California.
Scale is 4 micron per pixel, effective resolution is 6 microns, field of view is 4mm x 4mm.
Photographs thanks to Professor Nick Thomas.

The mass of the Beagle 2 instrument, not including the stepper motor for focusing, is 158 grams. The device will provide images of around six microns resolution, a magnification of about x 25 when brought up to within half an inch of sample surfaces. The microscope utilises the same digital camera system as the stereopanoramic cameras. It has a series of twelve light-emitting diodes, three each of wavelength equivalent to red, green and blue light for viewing samples plus a UV source to stimulate fluorescence.

Seen under the microscope, the spots on the Damien Hirst calibration target have a relief, providing an excellent way for the depth of field of the microscope to be tested. The surface of the aluminium plate has been left unpolished so there is a wealth of microscopic texture for checking the focus.

The microscope has several tasks. Firstly the study of rock surfaces both as they exist already on Mars and after removal of any weathering layer. A second application is to record the size and shape of particles separating out as atmospheric dust, which will help scientists to understand the scattering and radiative transfer properties of the martian atmosphere. Samples taken for analysis by other instruments will also be looked at by the microscope.

If Beagle 2 was to come across anything that looked like a fossil, and it is a very remote possibility, then the microscope will record the image.

Carbon on Mars

Darwin might have had two microscopes but it was really only his own eyes he used to search for fossils, just as fossil hunters today search chalk cliffs looking for ammonites. The main life detection instrument on board the Beagle 2 lander can look into rocks for the evidence of fossil organic matter with a greater chance of success than any microscope ever could. The Gas Analysis Package detects every atom of carbon in all its forms in samples that are combusted and oxidised over a series of temperatures to convert carbon and its compounds into carbon dioxide, a form that can be easily detected and measured. This is a very sophisticated blow pipe but the principle is the same.

One of the important jobs for the Beagle 2 mass spectrometer is to measure isotopic abundances. On Mars it will be trying to find organic carbon and carbonate minerals co-occurring, the former as the residue of living processes and the latter an

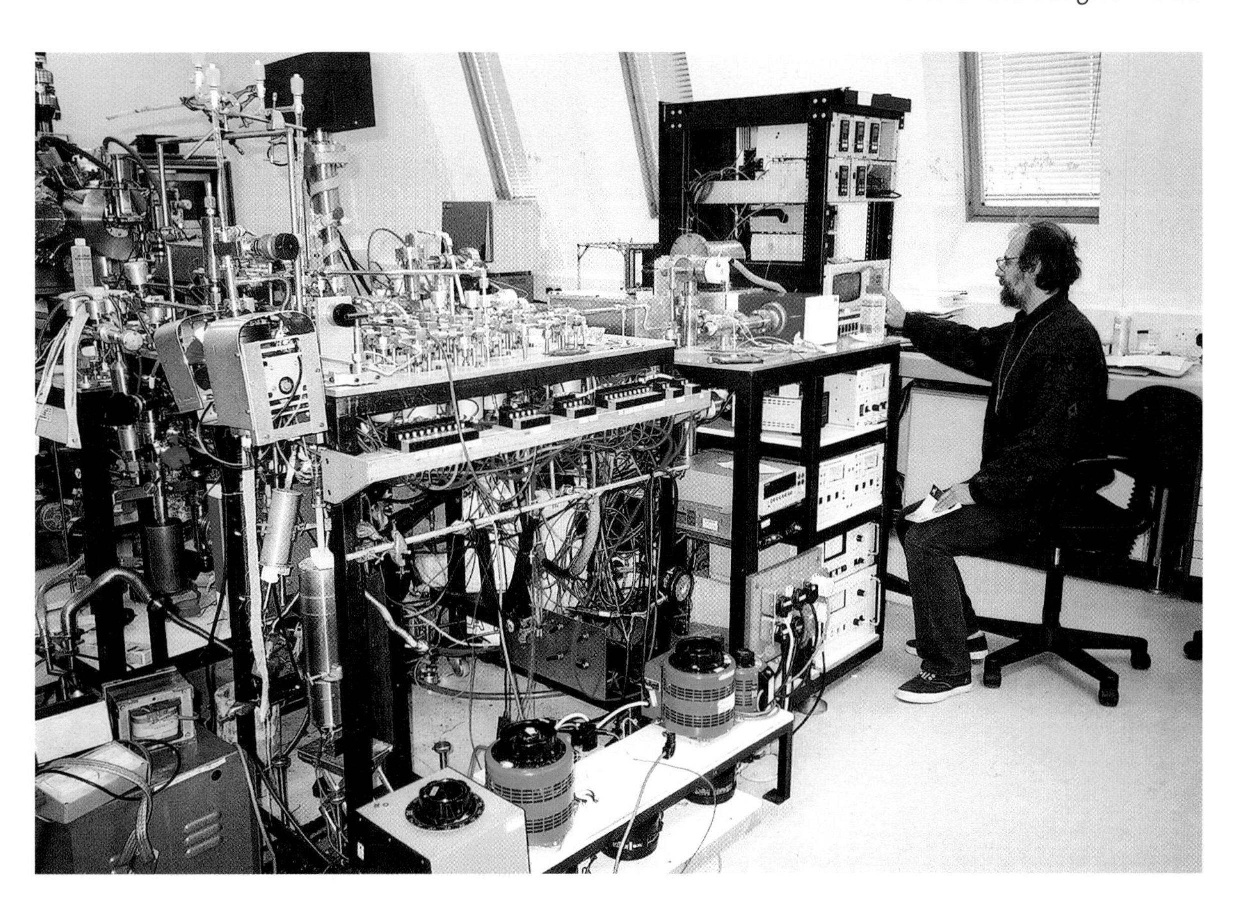

A mass spectrometer system used in the laboratory on Earth to carry out martian meteorite investigations.

indication of the existence of water, even if it is no longer present. The instrument will be used to detect any large-scale isotopic fractionation, the method that is used on Earth to show that biology has been continuously operating for nearly four billion years.

The search for water

The GAP mass spectrometer can also analyse samples for nitrogen and hydrogen, key elements in the biological cycle, the latter being part of water, which is the essential ingredient of life. The reason why searches for life begin by looking for water is that it is the universal solvent. Chemical reactions only work well if the reactants are intimately mixed together. Gases for instance readily combine but solids do not. Water allows biochemical molecules to come into close contact and undergo transformations essential for life processes. The vast majority of plant and animal life is water; the human body for example is eighty percent water acting as a solvent for biochemical reactions. Without water being present a search for life is likely to be in vain.

The GAP is a much compressed version of the laboratory instrumentation used to study martian meteorites. Instead of occupying several cubic metres it has been built to fit into the space the size of a shoe box.

GAP consists of five separate entities: First, a sample handling and distribution system (SHADS), twelve tiny ovens on a carousel, which can be rotated to allow each one to be used for gas extraction.

The gas is then passed to the second area, a gas processing unit, which is comprised of a sandwich of three titanium plates with galleries machined in them to allow the movement of gas from place to place. Screwed into the plates are thirty one tiny solenoid valves which direct the gas flow through a number of reactors. These allow the gas to be cleaned, converted, compressed and expanded according to the instrument's needs. Some of the reactors contain standard materials to calibrate the mass spectrometer on Mars.

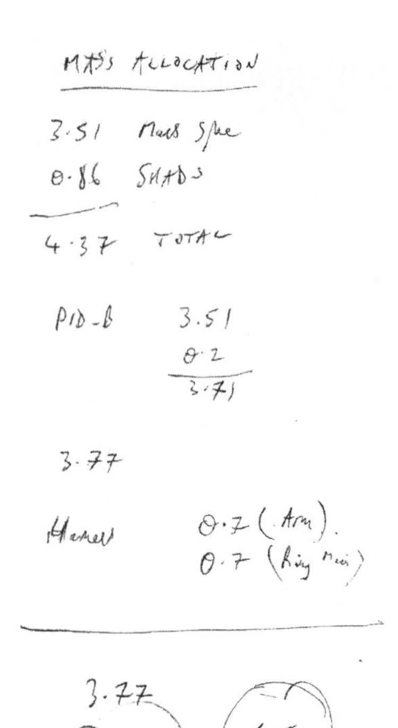

Mass budget for the gas analysis system, carefully calculated on a lunch-time napkin. With thanks to Dr Ian Wright.

The third part of the system, the mass spectrometer itself, is a single focusing magnetic sector instrument, which has several collectors at fixed focuses. Ions are produced by electron impact in a novel solid ceramic source designed so as to protect the delicate ion optics during the rigours of launch and landing. The voltage to accelerate ions can be scanned to record a mass spectrum at one of the collectors. The magnet is made from a special alloy giving a very strong magnetic field for a much smaller mass of metal than conventional iron magnets.

The fourth unit in the GAP is a pumping system of sorption and ion pumps which provide vacuum to various sections of the instrument.

Finally, as the fifth section, there are some GAP-specific electronics, a high voltage supply in a sealed box, some logic cards, which allow the various functions of the sample handling to be programmed, and an ion counting and amplification system.

The total weight of the GAP system is five and a half kilograms, not including parts of the electronics, which are common for all of Beagle 2, such as power generation and the central computer.

Working with instruments relying on magnetic fields means that stray interferences can cause problems. FitzRoy, worried that his precious chronometers were affected by iron

The Beagle 2 gas analysis package mounted on its construction plate just prior to being fitted into the lander. The instrument sent to Mars will carry out the functions of four different laboratory mass spectrometer systems within a space approximately the size of a shoe box.

cannons, swapped them for brass guns. The various components in Beagle 2 have been screened against strong magnetic fields from the GAP mass spectrometer.

Units

Balances for weighing samples do not lend themselves readily to space missions (or the motions of a ship). Estimates of the size of samples studied by Beagle 2 will be made on the basis of capacity of the sample containers (seeing how much gas is needed to occupy the space above the sample) and visual inspection by the camera of the material taken before analysis.

Charles Darwin was notorious for his eccentric measures of specimen weight. His primary unit was the extent to which his water bottle was filled (sort of equivalent to the gas method). For small adjustments he used bullets and pellets of shot. Hence the observation "large black rat – weighs flask with water, without bottom two bullets, four pellets".

Blowing in the wind

On Earth methane is a key to detecting current biological activity. Methane is unstable in the air but is still present at a level of around two parts in every million parts of air. A fine balancing act is taking place. Methane is destroyed by chemical reactions such as photolytic processes but it is continually replenished into the air as a result of biological reactions. Methane is produced continuously by many microorganisms such as those active in the rumens of cows, in submerged soils like paddy fields, marshes and peat bogs and even degrading rubbish in landfill sites.

If martian bacteria exist, surviving and growing, protected deep down from the harsh surface conditions, they are likely to have a simple metabolism producing methane. So detecting methane in the martian atmosphere at the Beagle 2 landing site would mean that somewhere, not necessarily close by, perhaps deeply buried under an icecap of martian permafrost, there is microbial activity.

The methods to make the measurements of methane, a greenhouse gas, have been perfected for the purpose of understanding global warming on Earth. The same procedures can easily detect any methane on Mars even if it is present at less than a few parts in a billion parts of the martian atmosphere.

A major terrestrial source of methane is produced below the surface in landfill sites by microorganisms degrading rubbish.

Carefully examining the atmosphere can provide information about not just biological life but the life of the planet. For example detection and measurement of the rare gas xenon can provide clues about the origin and history of the martian atmosphere and about what radioactive elements might have been present to heat the planet at the time of its formation.

Methane produced by bacterial activity in ruminant animals is belched into our air.

Methanogenesis

Extremophiles, organisms which can survive harsh conditions, have to be opportunistic in using what is available as an energy source. The simplest form of metabolism known about is the reduction of carbon dioxide to methane. There is plenty of carbon dioxide on Mars, the atmosphere is more than ninety-five percent of the gas, and so Beagle 2 will look for the product of methanogenesis in the hope of sniffing out life.

The age-old question

Beagle 2, like Darwin and all geologists, places great store on knowing the age of rocks and the timing of events. Therefore determination of the potassium content in martian rocks is very important. Potassium degrades over a very, very long, but known, timescale to argon, and the abundance of argon, a rare gas like xenon, will be measured in samples by the mass spectrometer. Thus it will be possible to estimate when rocks solidified from molten lava, the so-called crystallisation age. No one has ever dated a rock *in situ* on a planet (other

The GAP mass spectrometer which will be used to measure the abundance of argon and neon, in the hands of one of the design team.

than Earth!) before so, if Beagle 2 achieves this goal, it will be a first of considerable importance.

The gas analysis package, the central instrument in the search for signs of life, can also be used to determine the abundance of yet another rare gas, neon and its isotopes, in rocks. The amount of the mass-21 isotope of neon can reveal how long rocks have spent on the very surface of the planet (called the exposure age). This is important information in working out the processes which have modified the landing site. The study can be made because Mars has such a thin atmosphere that cosmic rays make it right down to the surface and cause nuclear reactions to occur in the rocks. This might be bad news for future astronauts but the cosmic rays afford the possibility of obtaining key geological information.

Mad about the weather

Admiral FitzRoy's grave.

One of the people most responsible for FitzRoy not being properly credited, for his contribution to Meteorology after his death, was Darwin's cousin, Francis Galton, who wrote a report denigrating the Admiral's weather forecasts.

By the west door of a very ordinary church in the South London district of Upper Norwood is a tombstone, surrounded by a picket fence. The inscription on the tomb, a bible quote about the fickleness of the wind, provides the identity of the owner – Admiral Robert FitzRoy – too often the forgotten man of HMS *Beagle*.

In truth FitzRoy is lucky to have a tomb at all. In death he committed a cardinal sin, taking his own life on 30th April 1865 by cutting his throat with his razor. His final act was in a way precipitated by the lack of recognition of his achievements. Not those aboard the *Beagle*, his seamanship, navigational skills and abilities as a surveyor were praised before Parliament by a grateful Hydrographer to the Navy, Francis Beaufort. But FitzRoy's subsequent careers as MP, Governor of New Zealand and statistician at the Meteorological Office go almost unmentioned. True the former two jobs ended in debacle; FitzRoy and another politician threatened to settle their differences with duelling pistols and as New Zealand's Governor let down by Government, he made a hash of land settlement. FitzRoy eventually was made an Admiral by the Navy and was elected a Fellow of The Royal Society. But as founder of the Meteorological Office, his contribution to understanding the weather, instead of being acclaimed, was often ridiculed.

It was whilst on board HMS *Beagle* that FitzRoy began to believe it was possible to foretell, or to use his favoured word, forecast, the weather. His critics, including *The Times* newspaper, which carried his forecasts, rather unkindly called them prophecy. But through keeping a watchful eye on atmospheric pressure and wind speed, FitzRoy never lost a man in a storm at sea. By following his own rhyming maxims, including the one many have heard:

> "Long foretold, long last
> Short notice, soon past"

FitzRoy was able to predict the coming tempest and take evasive action.

In 1855, following a conference on Meteorology at sea, held in Brussels, the British Government, the Board of Trade, recognised that it was economically prudent to record weather conditions throughout the world's oceans, where its merchant (and Royal) navies did their business. The Board sought the advice of the Royal Society and FitzRoy was recommended as statist (statistician). It is not clear what the Board had in mind as a use for the data it expected FitzRoy and his three helpers to collect. Presumably a rather unhelpful 'this is the sort of weather to expect here' message to sea captains.

FitzRoy had other ideas. It was of no use to him to receive a report from a ship arriving in the Thames if it had experienced a storm in the Channel. By the time anyone made the return trip the weather was gone. But the statist had a practical solution, he instigated reporting by the newly invented radiotelegraph, made possible by the dots and dashes code of Samuel Morse. And with immediate advice available to shipping, FitzRoy began to supply instruments for proper measurement rather than just rely on eye-balled observations. Soon FitzRoy had divided the sea around the coast of Britain into sectors some which we still know and love.

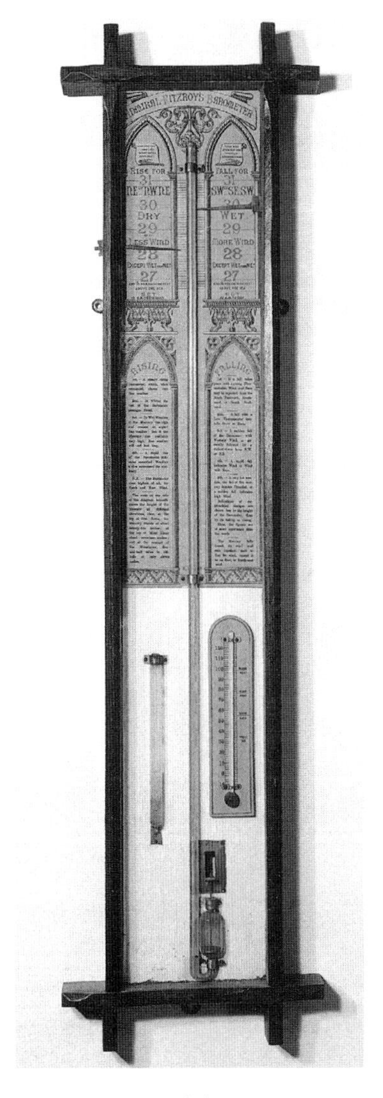

A barometer of the type FitzRoy advocated as an aid to sailors. Courtesy National Maritime Museum, London.

Epitaph on FitzRoy's grave

The wind goeth toward the south, and turneth about unto the north;
It whirleth about continually, and the wind returneth again according to his circuits.

Ecclesiastes 1.6

FitzRoy was not satisfied with gathering information; he saw a life saved or a journey made easier by advanced knowledge as common sense. Thus the Head of the Meteorological Office provided every port with a barometer of his own design for any prospective sailors to check conditions before putting to sea. This was an immediate success, so much so that FitzRoy became unofficial forecaster by Royal Appointment to Queen Victoria; a report for a channel crossing from Folkestone to Boulogne states: "Weather on Friday (March 4th, 1863) favourable for crossing – moderate, mild, cloudy, fine, perhaps showery at times."

Captain Robert FitzRoy wrote in the ship's log:
"Soon after one the sea had risen to a great height and I was anxiously watching the successive waves, when three huge rollers approached, whose size and steepness at once told me that our sea-boat, good as she was, would be sorely tried.
For a moment our position was critical; but like a cask, she rolled back again. Had another sea struck her, the little ship might have been numbered among the many of her class which have disappeared."

Blow the man down

Captain Beaufort, the Hydrographer for the Navy, was ultimately in charge of all the third *Beagle*'s voyages. He was christened Francis, his sister was also named Frances. Perhaps Beaufort went to sea at an early age to avoid the confusion. He became a desk-bound sailor after receiving sixteen wounds in one naval action alone and then another five. Maybe he was accident prone, anyway making maps was slightly less risky.

Although a brilliant map-maker Beaufort is more generally remembered for the Beaufort scale, an empirical system for reporting the wind strength at sea in the ship's log. Beaufort's scale was used by HMS *Beagle*, the first vessel to do so routinely. But the range of values from 1 to 12 was inadequate for the winds experienced off Cape Horn by the ship. FitzRoy, commenting to his superior in March 1833 said "You would have smiled at hearing some of my shipmates saying during the last cruise 'if Captain Beaufort were here now he would call this fifteen' ".

Although now superseded, Beaufort's notation can still be heard in the daily broadcasts of the shipping forecast made by the BBC.

'Sorely Tried' by John Chancellor who described the painting, based on the account of her Captain, as showing: "HMS Beagle *at 1.45pm on the afternoon of 13 January 1833, sixty miles WSW of Cape Horn, which she had rounded twenty-three days previously." With thanks to Mrs Chancellor for permission to reproduce.*

Off scale

The winds on Mars are ferocious, they reach 200 mph, however, because of the thinness of the atmosphere, they have much less force. They do whip the dust into the atmosphere as those 1909 astronomers, and many since, have observed.

Beagle 2 carries a wind sensor, mounted on the PAW, which operates on the principle of measuring the cooling effect of air passing over hot films. The differences between two films, one in the wind and the other a reference, can be used to calculate a two-dimensional vector. The sensor is calibrated accurately up to thirty metres per second although it can measure higher wind velocities with less precision. Thirty metres per second is only 108km/hour or 68mph, a measly storm force on the Beaufort scale, but hopefully Beagle 2 will not need the full range or the mission will end in disaster.

Wind information is not just scientifically interesting, it is needed to predict the power used from Beagle 2's battery, in keeping itself warm. Like information from other environmental sensors it is valuable when sampling the atmosphere to study compositions.

Landing hazard

The winds on Mars come into play in identifying a safe place to land. The projected landing site for Beagle 2 was changed by a few degrees of latitude and longitude when data from an orbiting satellite predicted there was a risk of high winds at the place initially chosen.

Turning up the pressure

So if FitzRoy had his way the barometer or atmospheric pressure sensor would be the last instrument Beagle 2 would choose to leave behind. Attempts have been made to fly pressure sensors on the ill-fated Mars 96 and NASA's Polar Lander, another mission that went absent without leave. A sensor was supplied to the 2001 NASA landing mission but this launch was cancelled before it even took off. The Netlander group which lost out to Beagle 2 would have had pressure sensors at all the stations. Hopefully Beagle 2's pressure-measuring device will provide data at long last.

The pressure sensor measures pressure absolutely by determining the capacitance changes induced by bending a

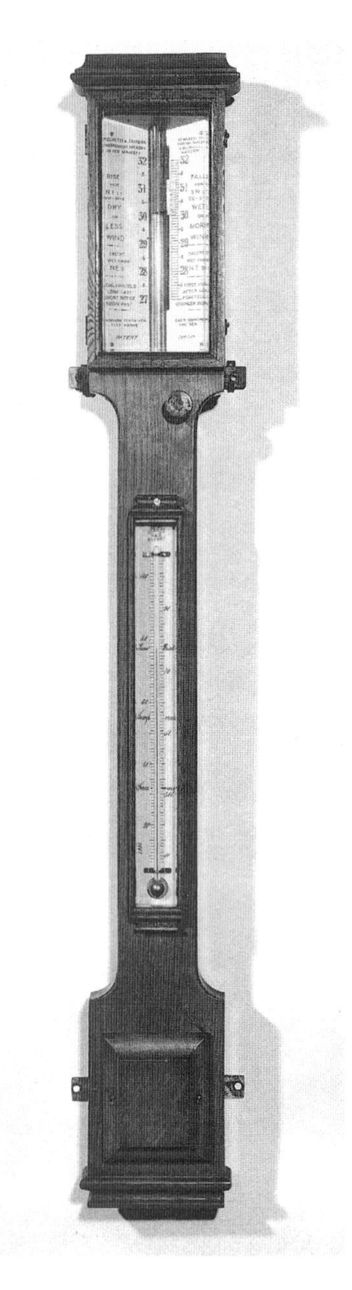

An anaeroid barometer carried by HMS Beagle *for measuring atmospheric pressure and used by Darwin to measure altitude above sea level. Courtesy National Maritime Museum, London.*

thin silicon diaphragm. It detects pressures as low as two millibars with a resolution of three microbars.

An important use of barometric pressure on Mars is to define the altitude of the Beagle 2 landing site. Darwin wrote to FitzRoy before leaving England asking if Beagle had a good set of mountain barometers. He was fascinated by Lyell's theories on uplift and said, "several great guns in the geological world have told me some points to ascertain which entirely depend on their relative height". Indeed, heights on Mars are arbitrarily measured against a 'sea-level' or martian datum defined by an atmospheric pressure of seven millibars. Beagle 2 will provide information about how far Isidis is below the datum.

Sunstroke

Beagle 2 has one environmental sensor that FitzRoy would not have been familiar with. Because the martian atmosphere is much more tenuous than that of the Earth and contains little or no protective ozone, harmful ultra violet, UV, light penetrates all the way to the surface. So Beagle 2 has a UV-measuring device. FitzRoy would not have known that the UV wavelengths of light existed.

One of the reasons given why NASA's Viking mission did not detect organic matter in 1976 may have been that the UV flux creates oxidising radicals in the atmosphere which destroy carbon-containing molecules. Beagle 2 will look underground whilst a UV sensor will provide the first measurement of the radiation flux at the martian surface.

Ice cold in Isidis

Beagle 2 lands in the spring in the Northern hemisphere of Mars so the night time temperature will hopefully only go down to -60°C. Beagle 2 has a number of temperature sensors, those for meteorological purposes are mounted on the PAW so it can be raised to about 0.6m into the air (to avoid heat output from lander), and the other as far away as possible on the outside edge of one of the solar panels. Both will measure air temperatures continuously in the range -10°C (a balmy spring day) to -60°C (the night time expectation) with an accuracy of ±0.01°C; outside this range the temperatures will still be measured but the accuracy will only be ±0.1°C.

Other temperature sensors are to be found on the mole which will allow a heat flow from the interior of Mars to be

The Viking lander searched for life only on the surface of Mars. Will Beagle 2 do better? First published by the Philadelphia Inquirer, *1976.*

measured, and within GAP to monitor oven and reactor temperatures. All the sensors are 0.3 millimetres in diameter platinum bead resistors. The information will be used regularly for non-scientific but nevertheless vital, purposes, that is to supply data for the Beagle 2 thermal model which allows the engineers to establish the health of the lander and

A martian environment simulation chamber used to test the functions of the mass spectrometer at low temperature.

control the thermal and power usage budget. This is an essential role in the running of the Beagle 2 operations. The priority will be to preserve the spacecraft during the night time hours and only do experiments when the power budget allows. Beagle 2 will have to be very carefully managed.

Alien invaders

Ships like HMS *Beagle* exploring in the mid-nineteenth century, carried live animals, amongst their stores, for meat. Being moored off the coast of Australia, one morning Captain John Lort Stokes, having enough for his own needs, released into the wild a dozen rabbits as "a provision for shipwrecked sailors". Stokes saw it as a good deed well done. Later generations of Australians would not thank him for the rabbit explosion, which they eventually tried to curtail by the introduction of another alien species, myxomatosis.

Protecting the planet

Beagle 2 will not be so irresponsible regarding contamination of Mars with organisms of Earth origin. In fact the lander has only be given the go-ahead to be launched because it has a known, verified, negligible load of microorganisms. In reality any organism present would be unlikely to survive the trip let alone colonise the surface of Mars. But no one is taking any chances, no one wants a role reversal of the scenario predicted in Wells' War of the Worlds. Hence the Planetary Protection rules, which cover both Earth and Mars, are taken very seriously by spacefaring nations.

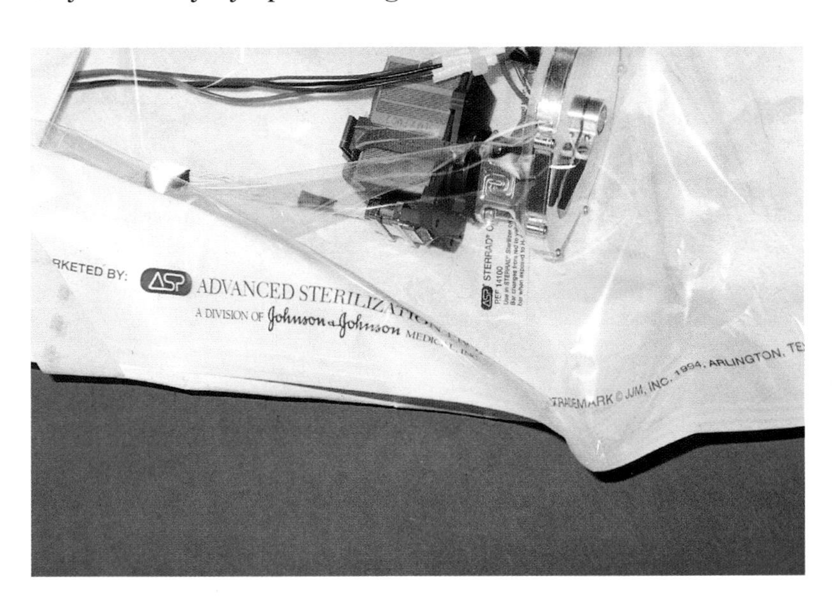

One of the methods used to sterilise Beagle 2's components was a plasma treatment more commonly used in hospitals to process heat-sensitive apparatus.

Planetary Protection

After the USSR launched the first Earth satellite in 1957, and thereby opened the space age, the International Council for Science, established its Committee on Space Research (COSPAR) to address issues of biological contamination being moved around the solar system. Policing planetary protection remains the responsibility of COSPAR.

Beagle 2 is classified as a category IVA mission, that is it will land on Mars but does not have the capability to detect actively metabolising organisms by methods which test viability in growth media. The regulations permit a bioload of not more than 300 spores (the most resistant form of microorganism) per square metre of lander, with a total limit of 300000 spores in total.

Beagle 2 was able to meet this requirement by submitting all components to sterilisation processes and then assembling the lander in the specially designed aseptic assembly facility.

Some scientists believe that forward contamination of Mars by landing spacecraft is so unlikely it does not warrant the additional costs imposed on a mission by stringent sterilisation. Others advocate far more stringent precautions will be needed if future missions land in those parts of Mars where there may be possible niches where terrestrial microbes could start to grow. But any experiments designed to detect minute traces of the chemistry of life, such as those of Beagle 2, clearly could be compromised by the presence of

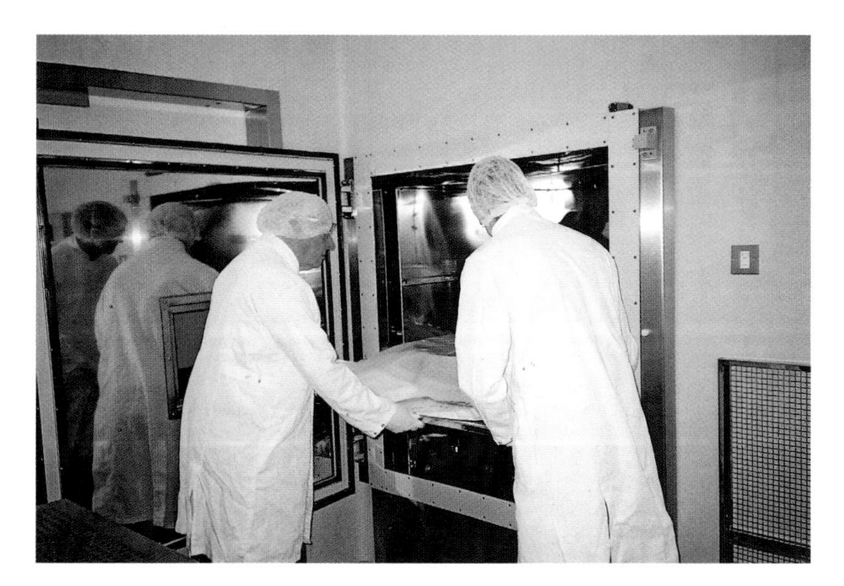

Other flight hardware could withstand sterilisation by high temperature for many hours.

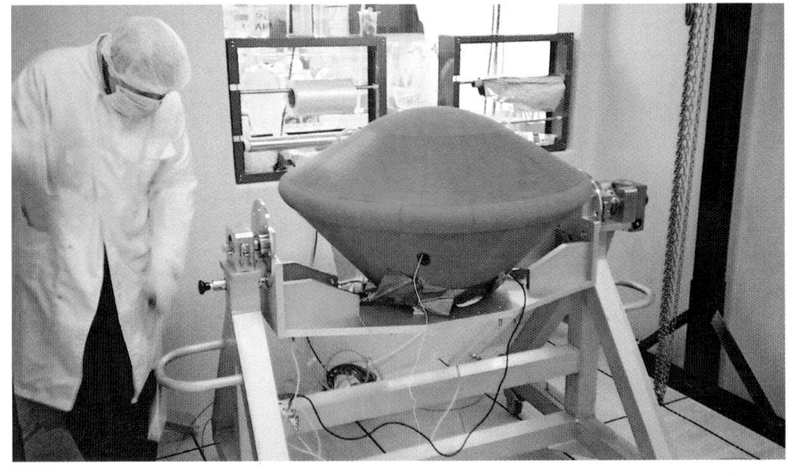

The probe, showing the front and back covers which prevent any contamination of those parts which will land on Mars.

terrestrial organisms, so even dead spores, or other carbon-containing debris, are removed.

Once the completed lander had been assembled and various parts of the landing system installed, the front cover and back cover were fitted together. A microbial filter allows gas pressures to be equilibrated during launch and space flight. The front cover acts as a heat shield against the fierce temperatures that will be encountered on entering the martian atmosphere; both provide a biological shield protecting the lander against contamination once out of the aseptic assembly facility and will themselves be self-sterilised by the entry temperatures.

Today the Galapagos Islands, known to thousands of tourists for its links with HMS *Beagle* and Charles Darwin, is fighting to remain a sanctuary for its indigenous creatures, stringently restricting invasion of any non-native species.

Beagle 2 goes to Mars in the knowledge that it lands as a responsible visitor.

Galapagos wildlife, courtesy The Galapagos Conservation Trust, London, photographer Chris McFarling.

The ship that sailed to Mars

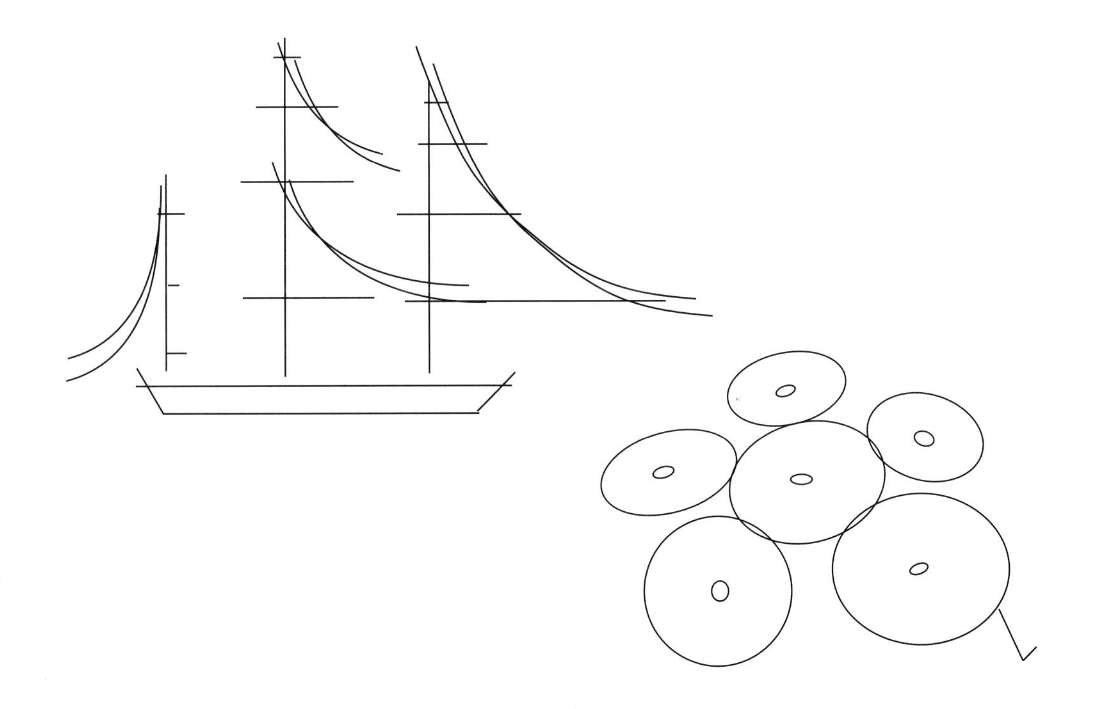

The last days in the career of HMS *Beagle* are remembered whilst another journey of discovery, to Mars, is just beginning some one hundred and seventy years later.

End of the voyage

Tiera del Fuego was not the most attractive part of the world in respect of climate. In a letter to his sister Caroline dated October 13th 1834, Darwin wrote "I suspect we shall pay T. del Fuego another visit; but of this the good Lord deliver us; it is kept very secret, lest the men should desert; everyone so hates the confounded country".

Even FitzRoy was affected, the thought of going back south without a companion ship (he had just received the Admiralty orders to dispose of *Adventure*), and not being able to complete the survey of the southern part of South America as he desired, made him as Darwin described "thin and unwell, this was accompanied by a morbid depression of spirits and a loss of all decision and resolution".

In consequence the Captain attempted to resign his command in favour of John Wickham. In reply the First Officer stated, "that when he took the command nothing should induce him to go to T del Fuego again". Wickham then asked the Captain what would be gained by his resignation. "Why not do the more useful part & return, as commanded by the Pacific. The Captain at last to enormous joy consented & the resignation was withdrawn."

The Beagle Channel photographed from space. It was discovered by FitzRoy on the ship's first voyage. NASA image.

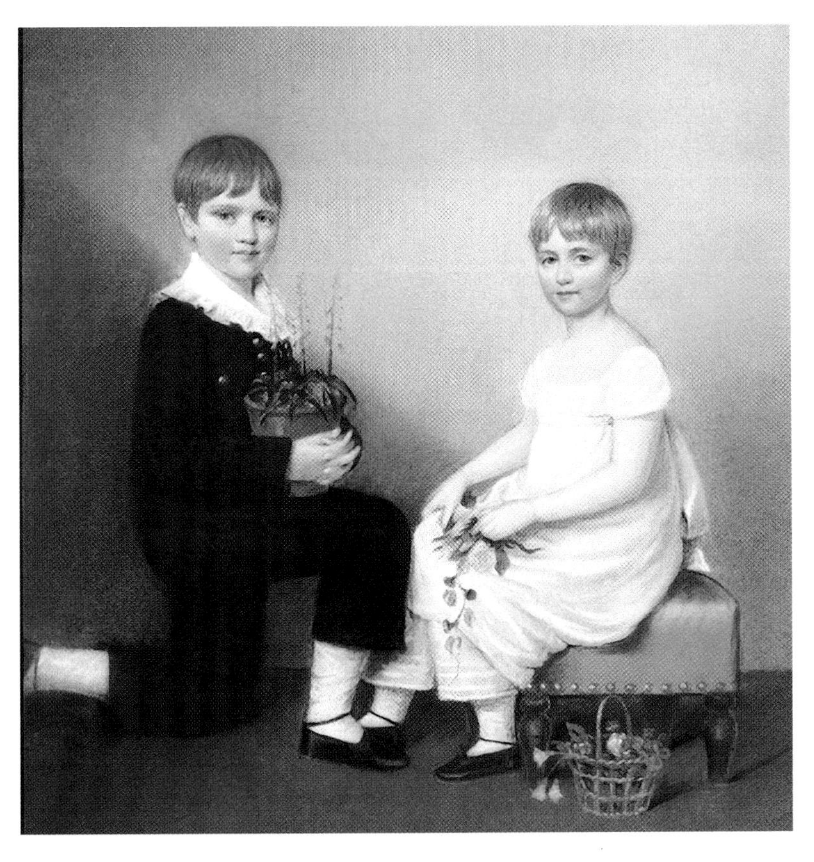

Charles Darwin (aged six) and sister Catherine by Sharples. Courtesy G.P. Darwin on behalf of Darwin Heirlooms Trust and English Heritage.

"Hurra, Hurra it is fixed the *Beagle* shall not go one mile South of C. Tres Montes," wrote Darwin to sister Catherine, November 8th 1834.

The second letter obviously did not arrive in time to allay the fears of the Darwin family because in response to the first missive, Charles was told by his father to take the next passage home.

When HMS *Beagle* did arrive at Devonport nearly two years later, Darwin hot-footed it to Shrewsbury leaving FitzRoy to take the ship to Greenwich and Syms Covington to cope with the precious collections.

How to win friends

At Greenwich news of the famous voyage had preceded HMS *Beagle* and even in 1836, PR was perceived to be important. Arrangements were made for visitors to be received. Dignitaries would board via the accommodation ladder to the poop deck whereas access to the general public would be granted by way of the main decks. Bartholomew Sulivan recalled events in his memoirs. Standing sentry in charge of

Sir Bartholomew Sulivan in his Admiral's uniform.

the companion way, Sulivan intercepted a "less than respectable" looking man, who, although he was accompanied by a pretty and stylishly dressed woman, was directed to the gang plank just three inches wide with a rope for hand holds. The pair turned out to be none other than the Astronomer Royal, George Biddle Airey, and his wife. FitzRoy was not best pleased. Despite his *faux pas* Sulivan was made Admiral and ultimately knighted. Airey is remembered for refusing to authorise a paper announcing the British discovery of Neptune, leaving the French to claim the honour.

Martian autumn

Whilst Beagle 2 will have plenty of media events in conjunction with launch and landing, it will not be coming back, so there should not be any similar embarrassing moment. That is unless some future Mars astronaut scoops the lander up and returns it to Earth. It's not an outrageous suggestion, Apollo 12 astronauts collected pieces of Surveyor VII during their lunar adventure.

Science fiction writer Steven Baxter, in his short story *Martian Autumn*, does indeed predict that Beagle 2 will be found and preserved under a translucent dome.

"The Beagle wasn't much to look at. It was just a pie-dish pod that had bounced from the sky under a system of parachutes and gas-bags. Disk-shaped solar panels had unfolded over the dirt, and a wand of sensors had stuck up like a periscope. When people had come looking for it five decades after its landing, Beagle had been all but buried by wind-blown toxic dust.

And yet, by taking its tiny soil samples and sniffing the air Beagle 2 had discovered life on Mars".

HMS *Beagle* is still talked about one hundred and seventy years after its journey. It would be gratifying if Beagle 2 stays in the memory for even fifty years.

In the first instance it is a matter of surviving for one hundred and eighty days. After arriving in the spring on Mars, Beagle 2 expects to be working throughout the summer and if it makes it to the martian autumn, it will be a bonus.

Putting Beagle 2 in the shadow

It should at least be possible to tell a future generation of astronauts visiting Mars where to look for Beagle 2.

The landing site could be anywhere in an ellipse about the size of England below the M4 motorway. Hardly the sort of information to give someone you are asking to drop in for a cup of tea. But there is an ingenious way of pinpointing the precise location of Beagle 2.

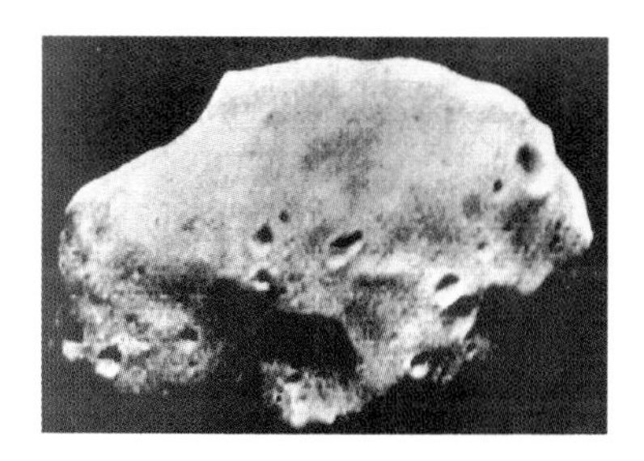

Phobos, one of the two moons of Mars, as seen by Mariner 9.

Fortunately it is possible to model the path of the shadow of the tiny martian moon, Phobos, over the landing site. When the UV sensors, part of the environmental monitoring package on-board Beagle 2, detect a reduction in flux caused by the shadow overhead, the coordinates of the lander can be calculated.

Around February 2004, a month or so after Beagle 2 reaches the surface of Mars, the shadow of Phobos will pass repeatedly over the vicinity of the landing site in the Isidis basin. The time the lander is in the shadow will allow computation of the position of Beagle with an accuracy of only a few kilometres.

Various instruments on the Mars Express orbiting platform will be able to focus their attention on the landing site to maximise the science return; for example the high resolution stereo camera (HRSC) and the geology mapping experiment (OMEGA) will have accurate ground-truth for at least one place on Mars. The planetary fourier spectrometer (PFS) experiment will collaborate on temperature and atmospheric measurements.

Gone but not forgotten

Most of Beagle 2's forerunners ended ignominiously as scrap. The third HMS *Beagle* was no exception. At the end of her last voyage in 1845 she was decommissioned and transferred to the

A typical scene on the Essex marshes where HMS Beagle ended her life as a Customs Watch Vessel.

By Order of the Lords Commissioners of the Admiralty.
FOR Public SALE, at Lloyd's Captains' Room, Royal Exchange, on Friday, May 13, 1870, at half-past 2 o'clock, H.M. paddle-wheel STEAM-VESSEL HYDRA, 818 tons B.M., built in H.M. Dockyard, Chatham, in 1838, copper-fastened and coppered, fitted with side lever engines by Watt, of 220-h.p. nominal, tubular boilers, with brass tubes from H.M. factory, Woolwich, in 1856, and feathering floats. Now lying at H.M. Dockyard, Sheerness.

H.M. paddle-wheel Steam-vessel Adder, 241 tons B.M., built at Harwich, copper-fastened and coppered, fitted with beam engines by Boulton and Watt, of 100-h.p., tubular boilers with brass tubes. This vessel is sold with her entire outfit. Now lying at H.M. Dockyard, Chatham.

H.M. sailing lighter Aid, 155 tons B.M., built in H.M. Dockyard, Chatham, in 1838, copper-fastened and coppered; is sloop rigged, carries a good cargo on a draught of 8 feet forward and 8.9 aft.; has been well kept up, and is sold with all her stores. Now lying in H.M. Dockyard, Chatham.

H.M. Coastguard watch-vessel Beagle, about 243 tons B.M., was built for 10-gun brig; copper-fastened and coppered. Now lying at Pagleshum, on the River Roche, near Southend.

H.M. Coastguard Cutter Lively, 100 tons B.M., built at Lymington, and copper-fastened, yawl-rigged, is of light draught of water, and very abundantly found in stores. About two years since she was repaired and re-coppered, is said to be sound and tight in the bottom, and is well adapted for a yacht, pilot-boat, or fer any trade requiring good sailing properties. Now lying in Harwich-harbour. For particulars, inventories, and orders to inspect, apply to

GEO. BAYLEY and WM. RIDLEY, 2, Cowper's-court, Cornhill.

Announcement of the sale of Watch Vessel Beagle *in* The Times, *by permission of the British Library.*

customs service, then a part of the Navy. She became the rather anonymous WV (watch vessel) 7. Her station was the mouth of Paglesham Pool on the river Roach in Essex. In midstream she served her purpose well but annoyed the local oyster catchers, who petitioned for her to be moved to the shore.

The sale of war-vessels ordered by the Lords Commissioners of the Admiralty was continued and concluded on Friday, with the following results:—Her Majesty's paddle-wheel steamship Hydra.—Tonnage (builder's measurement), 818; dimensions—length, 165ft.; breadth, 32ft. 10½in.; depth, 20ft. 4in.; built at Her Majesty's dockyard, Chatham, in 1838; fitted with a pair of beam engines, amidships, of 220-horse power; average speed, 8 knots—3,200*l*. Her Majesty's paddle-wheel steamship Adder.——Tonnage (builder's measurement), 241 tons; dimensions—length, 116ft. 4in.; breadth, 21ft. 4in.; depth, 12ft. 6in.; built at Harwich in 1826; fitted with a pair of beam engines of 100-horse power; speed, 9·24 on trial—1,100*l*. Her Majesty's Coastguard watch vessel Beagle.—About 243 tons B.M.; dimensions—length, 118ft.; breadth, 30ft. 6in.; depth, 12ft.; was built for a 10-gun brig, copper-fastened and coppered—540*l*. Her Majesty's sailing lighter Aid.—155 tons B.M.; dimensions—length, 68ft. 6in.; breadth, 23ft. 2¼in.; depth, 11ft.; built at Chatham in 1838;—bought in. Her Majesty's Coastguard cutter Lively.—100 tons (builder's measurement); built at Lymington and copper-fastened; about two years since she was repaired and re-coppered, and is said to be sound and tight—520*l*.

The ship was sold for five hundred and forty pounds, by permission of the British Library.

A last mention of her in customs records did not pinpoint the place but implied the owner of the land, Lady Olivia Sparrow, was owed the rent for her berth. There *Beagle* remained until 1870 when she was sold at auction to Messers Murray and Trainer for £540, presumably to be broken up.

Where those bits of fabled timber now reside is anybody's guess, perhaps one day we will have the answer.

A tiny relic of HMS Beagle, *a small box fashioned from the ship's timber. Courtesy National Maritime Museum, London.*

St Mars – the Patron Saint of poor space scientists?

Mars might have been the Roman god of war but Saint Mars seems to have been a man of almost entirely opposite character. Legend has it as a young man in the 6th century this Breton decided to devote himself to God, lead a solitary life, and, as a hermit, sleep on a stone bed. It could not have done him much harm because he was reputed to have lived until the age of 90 years.

A tiny church devoted to St Mars can be found in the hamlet of Marse, near Bais, southwest of Vitre in Brittany.

The saintliness of St Mars attracted a band of disciples over whom he presided as Abbot. The group divided their time between prayer and self-sufficiency. A story told about St Mars concerns a thief who tried to rob the hermitage of its meagre stock of food. Whilst trying to escape he was apprehended and dropped his booty. After helping the burglar gather up the scattered plunder, St Mars freed him with the blessing "go in peace and give up your evil ways".

Let's hope the underwriters of Beagle 2 are as forgiving if the promised sponsorship fails to materialise.

The church contains a beautiful stained-glass window depicting the hermit with his upturned palm held out. He deserves to become the patron saint of space researchers.